Conserving Nature in Culture

Conserving Nature in Culture
Case Studies from Southeast Asia

MICHAEL R. DOVE
PERCY E. SAJISE
AMITY A. DOOLITTLE

Editors

Monograph 54/Yale Southeast Asia Studies

Library of Congress Catalog Card Number: *2005-107533*
International Standard Book Number: *paper* *0-938692-82-8*
cloth *0-938692-83-6*

Distributor:
Yale University Southeast Asia Studies
P.O. Box 208206
New Haven, Connecticut 06520-8206
U.S.A.

Printed in U.S.A.

Contents

Figures

Tables

Appendix Tables

Contributors

Dr. Oekan S. Abdoellah, Director of the Institute of Ecology and Lecturer in the Department of Anthropology, Padjadjaran University, Bandung, Indonesia.

Dr. Andres Tomas Dans, Head Curator, Dr. Dioscoro S. Rabor Wildlife Collection, Museum of Natural History, University of the Philippines at Los Baños.

Dr. Amity A. Doolittle, Program Director, Tropical Resources Institute, and Lecturer, Yale School of Forestry and Environmental Studies, Yale University.

Dr. Michael R. Dove, Margaret K. Musser Professor of Social Ecology and Professor of Anthropology, Yale University.

Dr. Herminia A. Francisco, Associate Professor of Economics, College of Economics and Management, University of the Philippines at Los Baños, Laguna, Philippines, and Deputy Director of the Economy and Environment Program for Southeast Asia (EEPSEA), administered by the International Development Research Center (IRDC).

Dr. David Frossard, an independent anthropologist. He has taught for Semester at Sea; the Colorado School of Mines, Golden; the University of California, Irvine; and the International College of China Agricultural University.

Dr. Juan Carlos T. Gonzalez, Assistant Professor of Zoology, Institute of Biological Sciences, College of Arts and Sciences, and Curator for Birds at the Museum of Natural History, University of the Philippines at Los Baños, Laguna, Philippines.

Mr. Budhi Gunawan, Lecturer, Department of Anthropology and Researcher, Institute of Ecology, Padjadjaran University, Bandung, Indonesia.

DR. WILLIAM SM. GRUEZO, Professor of Botany, Plant Division, Institute of Biological Sciences, College of Arts and Sciences, University of the Philippines at Los Baños, Laguna, Philippines.

DR. LYE TUCK-PO, 2003–04 Quillian Visiting International Professor at Randolph-Macon Woman's College in Lynchburg, Virginia. Prior to that, she was attached to the Center for Environment, Technology, and Development, Malaysia.

MR. PAMPANG PARIKESIT, Lecturer at the Department of Biology and Researcher at the Institute of Ecology, Padjadjaran University, Bandung, Indonesia.

DR. RAJINDRA K. PURI, Lecturer in Social Anthropology and the convenor of the Masters Programme in Environmental Anthropology, University of Kent at Canterbury.

DR. PERCY E. SAJISE, Professor, School of Environmental Science and Management, University of the Philippines at Los Baños, Philippines, and Regional Director, International Plant Genetic Resources Institute, Serdang, Malaysia.

MS. MARILIZA V. TICSAY, Biodiversity Specialist, Biodiversity Research Programme for Development, SEAMEO SEARCA College, Laguna, Philippines.

DR. CLEOFE S. TORRES, Associate Professor of Development Communication, Department of Development Communication, College of Agriculture, University of the Philippines Los Baños, Laguna, Philippines

MR. VERNON VELASCO, Information Systems Analyst, Institute of Computer Science, College of Arts and Sciences, University of the Philippines at Los Baños, Laguna, Philippines.

MR. DANTE K. VERGARA, currently doing graduate work at Michigan State University and is University Research Associate, School of Environmental Science and Management, University of the Philippines at Los Baños, Laguna, Philippines.

Preface

In spite of a generation of heavily funded global efforts, there has been limited success in halting global environmental degradation and biodiversity loss. Recent reviews of the field attribute this to pervasive and fundamental failures in the focus of management and conservation programs (Whitten, Holmes, and McKinnon 2001; Clark 2002). We believe that one of the factors underlying these failures is systematic differences between the northern industrialized countries and the southern less industrialized countries in the way in which environmental problems are perceived and approached. Whereas there is great diversity within the North and the South, there also are systematic differences between the two in the way in which the major threats to the environment, and the best means of mitigating these threats, are framed. This also applies to the even more fundamental differences between North and South regarding the analytic frameworks or paradigms within which environmental threats are investigated. Scholars working in industrialized northern nations tend to naturalize their own scholarly norms and implicitly assume that they prevail, or at least should prevail, throughout the world. In fact, the scholarly conventions according to which the causes of and solutions to environmental degradation are framed, studied, and then reported, may vary profoundly between northern and southern nations.

This volume represents an effort to come to grips with this variation with respect to one region of the world, Southeast Asia. It represents the fruits of a four-year project funded by the John D. and Catherine T. MacArthur Foundation. The project was explicitly designed to be interdisciplinary and collaborative. Percy E. Sajise and

Michael R. Dove were coprincipal investigators on the grant. Dr. Sajise is a Cornell-trained biologist and Dr. Dove is a Stanford-trained anthropologist. Dr. Amity Doolittle, a Yale-trained social ecologist, joined the project at a later date to assist in analysis and publication of the project's findings. The project was formally coordinated by two institutions with wide experience in multinational research in the region: the Southeast Asian Ministers' Educational Organization Regional Center for Graduate Study and Research in Agriculture (SEAMEO-SEARCA, then headed by Dr. Sajise), the primary agricultural research and educational institution for the Association of Southeast Asian Nations (ASEAN); and the East-West Center (where Dr. Dove was then based), a federally funded U.S. organization that has carried out programs of education and research in the Asia-Pacific region since 1960. Much of the project's field research was carried out in partnership with two national centers: the Institute for Environmental Science and Management at the University of the Philippines at Los Baños (IESAM); and the Institute of Ecology (IOE) at Padjadjaran University in Bandung, Indonesia. These are two of the foremost institutions in the region carrying out research on the interface between society and environment.

Faculty and students at both IESAM and IOE, hailing from both the natural and social sciences, participated in the multiyear research program, as did a number of doctoral and postdoctoral students at the East-West Center and the University of Hawaii. (Altogether, the project's research findings served as the basis for one dozen undergraduate and graduate theses.) To foster a dialogue among all of the project participants, workshops were held once or twice a year throughout the life of the project, alternating in location between Los Baños and Bandung. A culminating writing workshop was held at the East-West Center, where project participants were brought together to write up the results of their work. Considerable project resources were devoted to translating and editing the writings of the contributors, making this for some their first international, English-language publication.

Throughout the project there was an emphasis on eliding the conventional concepts that contribute to the failure of so many conservation efforts. These include such things as the community bias: the persisting tendency of conservation projects to problematize local communities, in contravention to which the project supported research designed to identify both the ways in which local communities support conservation and the ways in which supracommunity forces undermine it. More notably, this refers to the unproblematized paradigms that govern not only the ways in which the environment and society are conceived, but also norms of scholarly research and representation, the nature of the scholarly voice, and the critical role of the scholar in society. Our project and this volume have endeavored to make paradigmatic differences explicit and to construct ways to talk across them but not erase them.

In this volume there is, as a result, a challenging mixture of views and voices. The volume reflects in this respect the reality of the diversity, contradiction, and even conflict within the global conservation community. We believe as editors that it is better to try to represent this diversity than to try to resolve it. For it seems likely that only approaches to conservation that recognize this human, social, and academic diversity as the reality, and indeed as a source of strength versus weakness, can have any hope of success.

Acknowledgments

THE PUBLICATION OF THIS VOLUME was made possible by the support given by a number of people and institutions during its long gestation. These especially include Marvel Kay Mansfield and James C. Scott at Yale's Program on Southeast Asia Studies, and Austin Zeiderman, who prepared the index. Institutions that have lent their support to this publication and/or the research upon which it is based include the East-West Center's Program on Environment, the SEAMEO Regional Center for Graduate Study and Research in Agriculture (SEARCA) in Los Baños, the Institute of Environmental Science and Management at the University of Philippines in Los Baños, the Institute of Ecology at Padjadjaran University in Bandung, Yale University's School of Forestry and Environmental Studies and Program on Southeast Asian Studies, and, especially, the John D. and Catherine T. MacArthur Foundation, which generously funded this project from start to finish by means of grants to Michael R. Dove and Percy E. Sajise. The staunch support of current and past MacArthur Foundation staff, notably including Kuswata Kartawinata, Dan Martin, and Judith Rayter, has been invaluable to the project editors and directors. The editors also wish to acknowledge the permission of the Island Press to publish here M. R. Dove's paper on intellectual property rights, an earlier version of which was published in *Valuing Local Knowledge: Indigenous People and Intellectual Property Rights*, edited by S. Brush and D. Stabinsky. None of the foregoing institutions are responsible for matters of fact and opinion published here, which are the sole responsibility of the joint editors and individual authors.

Introduction: The Problem of Conserving Nature in Cultural Landscapes

Michael R. Dove, Percy E. Sajise, and Amity A. Doolittle

> If we can give up this story of virgin wilderness ... we can lay to rest some of the misanthropy of old-fashioned conservationists and recognize that culture does not necessarily destroy nature.
>
> — Rebecca Solnit

Driven by increasingly rapid economic and political changes and prophecies of a great spasm of extinction unprecedented in the prehistory of the earth (Ehrlich and Wilson 1993), the problems of biodiversity loss and environmental degradation today occupy a central place in the work of both academics and policymakers (Turner et al. 1990; World Bank 1992; Peet and Watts 1996; Soulé 1991; Terborgh 1999). These concerns are particularly relevant in Southeast Asia where political and economic changes continue to accelerate at a rapid rate. Fundamental changes in social, political, and economic systems have had profound effects on the ways in which people use their natural resource base (Brookfield and Bryon 1993; Bryant and Parnwell 1996; Hirsch and Warren 1998). Throughout Southeast Asia in recent years, logging, monocropping, road construction, crop intensification, capital investment, planned settlements, and infrastructure and hydro-electric development have come to even the most marginal areas (Li 1999). All of these activities have raised fears of rapid degradation of the biophysical landscape and resource base.

The central finding of this book is that successfully stemming the tide of species extinction will require a radical reconception of the problem; that it will require not just coming up with answers to the questions being asked about conservation but asking different questions in the first place. The question that primarily concerns us in the following chapters is how the resource-use patterns of local communities bear on the issue of environmental conservation.

Rethinking the Problem

In most discussions concerning environmental conservation, human interactions with the environment are considered to be disturbances of the natural ecosystem that result in some sort of loss of integrity. As a result, the very existence and activities of the human population are often seen as harmful to the environment. Thus, the past approach to conservation has been to ask the questions: What are local communities doing that is inimical to the environment and how can we counter it? This book attempts to reframe the question by asking: What are these communities doing that supports environmental conservation and how can we strengthen, or at least not weaken, these activities (Dove 1993b)? Collectively, the body of work that follows refutes the notion that human manipulations of the landscape necessitate environmental degradation. In its place it is affirmed that human relations with the environment can not only sustain, but potentially even enhance, its integrity.

This reframing of the question concerning humankind's role in maintaining (as opposed to harming) critical dimensions of the environment builds on recent advances in both anthropological and ecological studies of tropical forests and their inhabitants. It was commonplace in the past to refer to "pristine" areas of the rain forest where "primeval" primary forests could be found, untouched by humans. But recent findings in ecology show that rainforests, like many ecosystems, do not remain productive when they are indefinitely in stasis; they require periodic human or natural disturbances to revitalize, maintain, or enhance ecosystem structure and function

(Anderson 1999). Indeed, the natural world is far more dynamic and changeable than scholars previously acknowledged; disturbance is necessary in many kinds of ecosystems, and change is the norm. Thus, the "balance of nature" metaphor has gradually been replaced by a new paradigm in ecology that stresses the nonequilibrium of ecosystems and the role of humans as an integral disturbance factor (Anderson 1999; Botkin and Sobel 1975; Pickett, Parker, and Fielder 1992; Scoones 1999; Zimmerer 2000). Recognizing the dynamism of the natural world challenges many of our popular understandings of the environment and conservation (see Cronon 1996).

In conjunction with a shifting paradigm in ecology, anthropologists are rethinking the role of humans in shaping the landscape. Recent findings in anthropology have shown that much of the tropical lowland forest—in Southeast Asia and elsewhere—is the product of many generations of selective human intervention and modification (both deliberate and inadvertent), optimizing its usefulness and at the same time often enhancing biodiversity (Ellen 1999, 137; Cf. Orlove and Brush 1996). The majority of the empirical work supporting these claims comes from the Amazon (e.g., Balée 1993, 1994; Roosevelt 1989; Raffles 2002; Denevan 1992) but there is emerging evidence that it also applies to Africa (Fairhead and Leach 1995, 1996) and a large part of Malaysia and the western Indonesian archipelago (Aumeeruddy and Bakels 1994; Dove 1983; Peluso and Padoch 1996; Ellen 1999; Puri, this volume).

Whereas this new understanding that disturbance, whether natural or anthropogenic in origin, is integral to a healthy environment has been embraced in academia, its role in policymaking is minimal at the moment. Following earlier efforts to integrate conservation programs with the needs and expectations of local communities (see Wells 1994; Wells and Brandon 1992; West and Brechin 1991; Western and Wright 1994), there has been a backlash and return in some quarters to a conservation paradigm modeled after the American national parks system, in which people are removed, at times even forcefully evicted, from the landscape in order to preserve areas that are considered unique or rich in biodiversity (Terborgh 1999). There are many reasons why this approach to conservation has been less than successful (Brechin et al. 2002). Forced evictions from traditional lands (Neumann 1992, 1996; Peluso 1993), increased pressures on a smaller natural resource base (Hirsch and Warren 1998; Peet and Watts 1996; Gunawan, Parikesit, and Abdoellah, this volume; Sajise et al., this volume) and the loss of access to valuable resources (Peluso 1992; Guha 1989; Neumann 1992) are a few of the reasons why local people may turn to unsustainable activities within the boundaries of parks and reserves—activities that often result in environmental degradation.

In the United States the removal of people from the landscape for biodiversity conservation through the formation of the national parks system has had, ironically, the unintended effect of reducing biodiversity in certain areas. As a result, it has been found that it is necessary to "reintroduce" human disturbances of the ecosystems—disturbances that had been either banned or rendered obsolete through changes in the political economy of the region—in order to maintain historic levels of biodiversity. An example is the reversal in fire policy of the U.S. National Parks Service in Yosemite National Park. By the late 1960s, foresters in Yosemite began to wonder why the sequoias *(Sequoiadendron giganteum)* in the park were not regenerating, why incense cedars *(Celocedrus decurrens)* were flourishing in the Yosemite Valley, and why many of the park's vistas and open woodlands had vanished to young cedars—including more than half of the meadowland in the valley. It was discovered that sequoias rely on disturbance for regeneration—the heat from forest fires allows the cones to fall and burst and release their seeds. Without fires, these giants could not regenerate, and as a result the incense cedars could establish themselves. Furthermore, a fire history of the region determined that the longest interval between fires prior to the 1800s was fifteen years. The historic management policy of fire suppression (essentially a policy of removing not only natural but also anthropogenic disturbance) had resulted in alteration of the landscape of the national park to such an extent that it no longer resembled the "pristine" ecosystem that the park service had set out to preserve. Finally, in the 1970s controlled burns were initiated (see Solnit 1994, 294–308, for a treatment of fire in the American landscape). Now a plaque in a restored meadow in Yosemite reads:

> Two hundred years ago the Valley's meadows were much more extensive. Oak groves like the one across the way were larger and healthier. By setting fire to the meadows, and allowing natural fires to burn unchecked, the Valley's Native American inhabitants burned out the oak's competitors and kept down the underbrush for clearer shots at deer. With leaf litter burned away, it was easier to gather acorns—the Indians' main food source. Without fires incense cedars are encroaching on the left side of the meadows and beginning to shade out the oaks. But now with controlled fires the NPS is reintroducing a natural process.
>
> (Quoted in Solnit 1994, 301–2)

The fact that the park service refers to their effort to produce controlled burns as reintroducing a "natural" process signals how difficult is it for conservationists to recognize human disturbance, such as Native Americans' use of fire, as a beneficial and appropriate type of disturbance within the boundaries of a national park. By calling the controlled burns part of a "natural" process, the park service can foster the impression that they are preserving a natural process rather than intervening in the evolution of the landscape (302). Despite the reluctance of the park service to embrace anthropogenic disturbance as a healthy part of ecosystem functioning, its experience in Yosemite demonstrates that human disturbances can play a critical role in maintaining and even enhancing desired environmental conditions and suggests that a close examination of the current conservation paradigm is necessary.

One of the aims of this book is to reconsider current thinking about biodiversity loss. The problems it addresses can be disaggregated as follows. First, there has been a tendency in the past to reify the concept of biodiversity, to identify it with the environment, whereas it is really just one measure of the character of the environment or one measure of the character of human relations with the environment. Thus, it misses the point to attempt to preserve biodiversity per se; what must be preserved are the conditions that sustain it. Second, we do not know enough about these conditions, and what we do know focuses on what diminishes biodiversity as opposed to what maintains it, yet the conservation of biodiversity depends not only on mitigating the former but on supporting the latter. We need to ask why some societies and not others can preserve biodiversity. What is the character of the relationship between environment and society, at both micro and macrolevels, that explains this? Third, does an abundance of species truly improve the health of ecosystems? Is there a clear, causative relationship between diversity and ecosystem functioning? Is it more vital to focus our conservation efforts on the sheer number of species or on other factors such as functional biodiversity (Wardle et al. 2000)? And, finally, how do we combine the approaches of the biological and social sciences in order to answer these questions (Nyhus et al. 2002)? How do we link the concerns of the researcher and the policymaker?

Expectations and Findings of this Book

The following chapters attempt to transcend the current confines of conservation debates in three ways. First, all of the chapters identify not just the social factors that threaten conservation but the social factors that support it. Second, the case studies all look beyond the local ecosystem and communities to the broader society for challenges to conservation. And, finally, these studies approach the challenge to conservation as a problem not of the environment alone but also of society. Much development and conservation planning in Southeast Asia and elsewhere has been characterized by an artificial dichotomization of the problems of conservation and the problems of society. An example of the quandaries to which this leads is the characterization of the conservation impasse between "people and parks." Political and economic reasons have made it increasingly impractical to separate or remove people from the landscapes and species that we are trying to conserve. This dichotomy between humans and nature may be conceptually flawed as well; Cronon (1996) has argued that any actions that increase the "otherness" of nature, or the separation of people from nature, increases its vulnerability. According to this argument, it is more productive to view people as an integral and perhaps even beneficial part of ecosystem functioning than as alien elements responsible only for its destruction. Likewise, it is better to think beyond the bipolar oppositions of natural and unnatural and embrace the full continuum of a landscape that is both natural and cultural.

Avoidance of orthodox dichotomies has the potential to resolve some of the most troublesome questions in conservation—such as how to "compensate" local communities for conservation interventions or how to "balance" conservation against economic development—by suggesting that these are the wrong questions to ask. If conservation is indeed associated with a particular pattern of social relations, then it is maintenance of these relations that will ensure that conservation succeeds. Attempts to achieve conservation without regard for these social relations are thus likely to be problematic.

The goals of this book are to help bring conservation science out of the reserve—in the sense of focusing part of its attention on the broader, nonbiological factors that affect conservation issues—and bring into the reserve the policy-oriented social sciences—in the sense of focusing some of their attention on the local, biological consequences, both costs and benefits, of the phenomena they study. Thus, this book builds on some of the earlier thinking about community-based conservation programs, which have sought to integrate the economic and livelihood needs of local communities with the goals of conservation of biodiversity (Wells 1994; Wells and Brandon 1992; West and Brechin 1991; Western and Wright 1994). But the findings of this book suggest that community-based conservation efforts are most likely to succeed if researchers and policymakers look for existing local practices which have a conservation function, rather then trying to bring in new activities and concepts from the outside.

The emphasis in the following chapters is on asking where, in the interaction between society and the environment across space and time, conservation is being most successfully and least successfully achieved. Four significant findings emerge from this approach to the problem: (1) humans in Southeast Asia have a long history of manipulating the environment in ways that enhance the conservation of useful and needed natural resources; (2) large-scale, state-imposed interventions often fail to reach the intended goals of conservation; (3) high levels of biodiversity can be found in many traditional agricultural landscapes; and (4) resource use at the local level is influenced by social, economic, and political events at the broader, regional level.

Several associated themes run through the chapters, the first of which is space. According to older theories of conservation, it is assumed that biodiversity (e.g.) is highest in areas furthest removed from anthropogenic disturbances and closest to seemingly undisturbed habitats. This assumption drives the management of many protected areas in which planners try to enclose the most "natural," species–rich areas in order to minimize human influences and thereby protect as much biodiversity as possible. The findings of this book suggest that the relationship between human society and biodiversity maintenance is more complicated than this. In fact, some of the following chapters demonstrate that there is a middle zone between the most anthropogenic landscapes (such as urban centers) and the least anthropogenic ones (such as undisturbed forests) in which high levels of biodiversity are obtained through human manipulation of the environment. In short, we can often identify a threshold of human activity at which some disturbance can actually enhance biodiversity while either less or more disturbance can diminish it. This finding potentially has great implications for the design of protected areas, given that it raises the possibility that a protected area's buffer zone may contain higher levels of biodiversity than the protected area itself.

The second factor involved in many of our discussions is time: there are distinct patterns of variation in environmental conditions over time. A number of scholars have suggested that relations between society and the environment around the globe follow a customary developmental path. Thus, Ludwig, Hilborn, and Walters (1993) suggest that all resource degradation follows a historically established pattern, which we must study and understand to have any hope of changing it. One of the most powerful means of analyzing this pattern involves the concept of "national environmental transitions." This has been the subject of analyses by Panayotou (1994) and Mather (1990): the latter argues that in the course of their economic development all nations go through a "forest transition" in which they first degrade and then restore their forests; the former argues that in fact this transition involves all natural resources. This correlation between the early phases of socioeconomic growth in a nation and environmental degradation is in accord with one of the fundamental findings of this study, namely, that there is an association between the dynamics and integrity of a society and the dynamics and integrity of the environment.

The analyses of Mather and Panayotou offer a number of insights into the wider spatial and temporal context that frames the conditions of environmental conservation. Their work shows us, first, that the conditions of biodiversity maintenance that we study at the level of the household, community, and even watershed, river system, or mountain are subsumed within—and thus affected by—larger national and international-level forces. Their work shows us, second, that studies of conservation at these various levels give us a "snapshot" of these processes and that we need to see how these snapshots fit into the longer temporal processes that precede as well as follow them. Thus, we need to know not just where Mount Makiling (see chapters 5, 6, and 7) and the Citarum river basin (see chapter 3) are today in terms of conservation, we need to know where they have been and where they will be in the future.

The case studies in this volume explore the nuances of human manipulations of the landscape, the cosmology of resource use, measures of biodiversity in human–manipulated landscapes, the impact of government interventions in resource use, and grassroots initiatives intended to reclaim local control over the maintenance of biodiversity. Issues of ecological theory, anthropological theory, and conservation policy are all explored. The remainder of this introduction is devoted to summarizing the four main sections into which this book is divided.

Historical Perspectives on Biodiversity and Society

In chapters 1 and 2, Puri and Lye illustrate the historical depth of human manipulation of the environment in ways that maintain and in some cases enhance the biodiversity of useful natural resources. Through in-depth ethnographic studies, they look at Malaysian and Indonesian societies that are traditionally referred to as hunters-gatherers, a term that usually evokes images of simple activities whose effects on the environment are limited by low populations and limited technologies, as though hunter-gatherers simply skim off the surplus of an ecosystem that they leave largely unaffected (Solnit 1994, 304). Puri demonstrates that the Penan not only reap the bounty of the land but also help manage forest biodiversity. He analyzes the role of the Penan in the creation and maintenance of forest vegetation communities. He concludes that the Penan manipulate the distribution of fruit trees, sago palms, and grasslands and retard natural forest succession through burning, pruning, felling, and the reoccupation of abandoned agricultural and settlement sites.

Lye's analysis of Batek cosmology and cultural practices complements Puri's more materialistic analysis of Penan resource use practices. Lye argues that the Batek consider human involvement in and knowledge and management of the natural world to be critically important in staving off ritual danger and preventing ecological collapse. Humans are thus perceived in indigenous terms to be an integral part of ecosystem functioning and maintenance. To the Batek, the historical depth to their knowledge of the environment, their understanding of how to use and navigate it, and their ability to transfer their knowledge about it to their children, makes separation from it highly problematic. To the Batek, the Western conservation paradigm, in which people are removed from an ecosystem in order to protect it is illogical. The way in which people conceptualize their natural environment indeed depends on how they use it, how they transform it, and how they invest knowledge in it through their actions (cf. Ellen 1999, 139). Given the fundamental difference between how people like the Batek conceptualize the relationship between nature and society and how Western or Western-educated managers may conceptualize it, it is not surprising that the Western model of exclusionary national parks has often failed in non-Western parts of the world.

Failures of State Conservation and Local Responses

Puri and Lye provide us with greater awareness of the potentially beneficial nature of small-scale human manipulations of the environment. This awareness leads us to the second focus of this study, why large-scale, state-imposed conservation plans so often fail to promote conservation. This occurs not only because small-scale human manipulations of the landscape may be critical to conserving resources but also because large–scale, technocratic and centralized state schemes that attempt to control people and resources often proceed without adequate understanding of the local social relations surrounding resource use (see Scott 1998; with specific reference to the environment in Southeast Asia, see Hirsch and Warren 1998). In fact, governmental interventions aimed at conservation often have the unintended effect of degrading the landscape, its biodiversity, and the social institutions that may have promoted conservation.

Many of the studies in this volume support this finding. For instance, in chapter 3, Gunawan, Parikesit, and Abdoellah analyze a number of agricultural development initiatives introduced into villages on the borders of protected forest in the upper Citarum river basin of West Java. Their study begins with an overview of the regional landscape and the socioeconomic conditions of the villages in the river basin. The authors then describe government agricultural policies and dairy industry initiatives that were designed to reduce pressure on the state forests. Paradoxically, these programs had the opposite effect due to a lack of coordination between government agencies and a lack of knowledge of the local socioeconomic needs and complex land use systems. This chapter demonstrates that the linkages between policy centers and peripheries, between broad state agendas and local village needs, need to be more clearly articulated in order to understand the impact of central policies on both the local people and the resources on which they depend.

In chapter 4 Frossard focuses on a peasant movement in the Philippines that has attempted to regain local control over seed diversity. This movement grew out of farmers' realization that, despite the promises of the Green Revolution and the work of the Philippine-based International Rice Research Institute (IRRI), no single super, high-yielding rice can satisfy all social, economic, and ecological conditions. On the contrary, rice yields that satisfy farmers' requirements are dependent on use of a myriad of varieties tailored to local conditions, and only farmers are in the position to know the nature of these. Frossard shows that, paradoxically, as a direct result of the scientific community's efforts to increase the productivity of rice cultivation, diversity of rice varieties in the field has decreased, at the same time as the collections of varietal specimens in ex situ storage at places like IRRI have increased. Indeed, Frossard suggests the international crop breeding institutions are the single major cause of the decline of in situ genetic diversity around the world. In a direct response to the impact of the crop breeding establishment on crop diversity, a local farmers' movement in the Philippines has attempted to restore rice diversity in the field and return control over the genetic diversity of rice varieties to the local farmers.

Although the dynamics surrounding conservation or control over natural resources often serve as a legitimizing tool for vested interests and state control (Hirsch and Warren 1998), Frossard shows how civil society can regain some of this lost ground through grassroots initiatives. Just as improved seeds become a means for dominating people and nature (Yapa 1996), the struggle to regain control of rice germplasm in the Philippines reflects the larger struggle by farmers to defend their social, political, economic, and cultural place. There is often a parallel between local contests over seed production and diversity and wider contests over political and economic power. In countries in which institutions of open political opposition have not been permitted to develop, movements such as the one described by Frossard have succeeded in creating alternative sites for the expression of contested positions. In many parts of Southeast Asia a wider range of local voices are heard now than was true even in the early 1990s, as a direct result of civil unrest over access to natural resources (see Hirsch and Warren 1998). History shows why the link between access to resources and political power is such a critical factor in the success or failure of conservation strategies.

Biodiversity in Traditional Agricultural Ecosystems

The third theme of these studies is the high levels of biodiversity that are found in certain agricultural landscapes. Agricultural biodiversity is primarily determined by the heterogeneity of the agricultural landscape, which is in accord with the recent work by conservation biologists and ecologists on the importance of ecological disturbances and patch dynamics in the maintenance of biodiversity. The studies in this volume contribute to a burgeoning literature on the subject of biodiversity on agricultural landscapes (Brookfield 1999; Brookfield and Stocking 1999; Brookfield et al. 2002; Collins and Qualset 1999; Schroth et al. 2004; Wood and Lenné 1999).

In chapter 5, Ticsay examines a variety of local agricultural practices that enhance biodiversity among the famous terraced hillsides of the Ifugao. Ticsay demonstrates how the traditional land use system of the Ifugao produces an uninterrupted and mixed flow of products to the household through the creation of diverse "patches" in the landscape. Analyzed individually, these patches appear specialized, monocultural, and generally low in biodiversity. For instance, coffee plantations, rice terraces and bean gardens are each specialized land uses; their integration into a wider household subsistence system with swiddens, woodlots, and home gardens provides a great variety of resources and contributes to high overall levels of biodiversity. The traditional resource management system has been very sustainable throughout the years and quite successful in maintaining economically productive and biologically diverse landscapes. What remains to be seen is how long this system will continue to be successful as new external pressures, such as changing market forces, affect household decision making.

In chapter 6, Vergara analyzes biodiversity on Mount Makiling in the Philippines in relation to one particularly important determinant: the fragmentation or patchiness of the landscape. The results of his investigation generally indicate that biodiversity increases as human-created fragmentation increases, up to some fixed limit, after which it begins to decrease. Vergara's study is particularly concerned with agricultural activities both bordering and within state forest reserves. He concludes that in order to manage forest reserves with proximate human settlements so as to maximize biodiversity conservation, a major portion of the forest reserve should be left intact, and regularly shaped (viz., like a square or circle), to act as a biological resource pool. Farmers' fields should be distributed in adjacent buffer zones in a pattern of sparse, discontiguous, and clustered clearings. Clearing large, contiguous tracts of forest should be avoided. Preferably agricultural clearings should be, again, regular in shape, square or circular, as opposed to being long and thin or having other complex shapes. Studies like Vergara's are critical for designing conservation areas that accommodate the needs of forest conservation as well as the needs of farmers in adjacent communities. Rather than seeing farmers simply as threats to forest biodiversity, farmers and conservationists can work together to design agricultural landscapes that complement the goals of both conservation and agricultural productivity.

Regional and Global Perspectives

In chapter 7 the regional dimensions of biodiversity maintenance are explored by Sajise and his colleagues using, again, the case study of the Philippines' Mount Makiling Forest Reserve and the agricultural communities that lie both within the reserve and adjacent to it (cf. Sheil et al. 2002). One of the most significant findings of the study is a positive correlation between an increase in farm and household size, on one hand, and greater on-farm biodiversity on the other hand. Conversely, a negative correlation is found between increased wealth and education and crop diversity. These findings suggest that farmers with more labor and land resources tend to diversify their crops, whereas farmers with more education and financial resources tend to be more profit-oriented and more inclined to take on the risks inherent in monocultures.

These findings also highlight the paradoxical consequences that conflicting state policies may bring about at the landscape level. In the Mount Makiling case, some state programs threaten agricultural diversity through policies aimed at market-oriented, monocultural development; at the same time as other programs attempt to conserve forest-based biodiversity by prohibiting all human activities within the forest reserve. Yet it is these same subsistence uses of forest resources that allows farmers to maintain highly diverse, non-market-oriented gardens. Thus, when the subsistence use of forest resources is combined with a highly diverse cultivation strategy outside the forest, we see positive consequences for both society and environment at the landscape level. This raises the question whether continued local use of forest resources, when viewed as part of a larger system of diverse agroforestry practices, is in practice more supportive of conservation goals than is the forced exclusion of people from the forest coupled with the introduction of intensive market-oriented monocultures. This raises questions, in short, about the "accepted wisdom" that the best way to conserve forests is to divorce local communities from them through increased market involvement.

In chapter 8, Dove argues that if conservation is indeed associated with a particular pattern of social relations then attempts to promote conservation that target only local people, without regard to what is happening in the rest of society, are likely to be problematic. He explains, in particular, the premises and problems associated with acknowledging indigenous resource uses and rights through monetary compensation. Dove's chapter begins by examining two serious obstacles to the use of intellectual property rights as a mechanism to address the issues of environmental degradation and inequity: first, compensation for intellectual property rights is likely to fail to reach the indigenous people responsible for biodiversity conservation; and second, if compensation does reach them, it is likely to change them and undermine the basis for conservation in the process. Dove argues that attempts to protect global biodiversity through the use of intellectual property mechanisms may, counterintuitively, ultimately lead to local impoverishment of biological resources unless they directly address the implications for these resources of the wider marginalizing, political–economic relationships in which they are located. Dove concludes his chapter by suggesting that local people and resources alike might be better off if policymakers, instead of promulgating new policies, would examine the existing national and international policies that are inimical to biodiversity conservation.

Conclusions

All of the studies in this volume highlight the importance of differences in local, national, and international conceptions of natural resources and their use and conservation. Lye's chapter on the worldview of the Batek explores the difference between the Western scientific view of conservation and the Batek's knowledge and understanding of the natural world. She examines the inherent contradictions within the Western model of national parks, which promotes both conservation of biodiversity and economic development, usually through tourism. She suggests that the traditional Western approach to protected area management, in which all of the people are removed from the landscape (except for tourism), could be informed with the approach of the Batek, who believe that active, local human management of resources is crucial to ecosystem health.

Common to the analyses of both Lye and Frossard is a critique of the notion that knowledge must emanate from research institutions or policy-making centers, as opposed to the people living closest to the natural world. Frossard explores this issue of "scientism"—the privileging of one particular form of scientific discourse over alternative ways of seeing—in the context of attempts by Filipino farmers to regain control over the reproduction of rice diversity in the field. The privileging of certain sources of knowledge is associated with the existence of a ubiquitous power imbalance between those who make conservation policies (and tend to benefit from those policies) and those whose livelihoods are affected (often negatively) by them. There is a corollary distinction between those who benefit from resource use politics and those who bear the costs of both conservation and degradation (see inter alia, Dove, this volume; Peluso 1992; Guha 1989; Hirsch and Warren 1998; Li 1999).

These studies all demonstrate that environmental degradation is not the necessary result of human disturbances; conditions do exist and can be fostered under which societies can support conservation. These studies also show that local people are rarely solely responsible for degradation; the impacts of national elites and their resource use polices are typically of equal if not greater importance.

The studies in this volume build on recent advances in ecology that acknowledge the need for disturbances to regenerate and revitalize ecosystems as well as anthropological studies that critically interrogate simplistic conceptual divisions between nature and culture. They point toward a conservation paradigm in which agricultural extension agencies, conservation agencies, and farmers would all work together to develop alternative paths for agricultural development based on diversified traditional land use patterns rather than intensive monocultures. One need only reflect on the slow but steady growth of small, organic farms in local communities throughout the United States to realize that alternative routes to modernity are indeed possible.

The essays in this volume make a number of contributions to the study of society and environment. First, they offer new methodologies for analyzing the relationship between society and biodiversity. Vergara has developed a novel fragmentation index that allows for a more accurate assessment of the effects of landscape patchiness on biodiversity and endemism. This fragmentation index promises to facilitate future collaborations among farmers, agricultural extension workers, and conservationists by offering new ways to determine optimal mixes of agriculture, agroforestry, and forest reserves. Other methodologies, developed in the chapters by Ticsay and Sajise et al., focus on the rapid assessment of environmental conditions through techniques such as vegetation and mapping profiles, kitchen and market surveys, participatory transect walks, and interviews with local people. These methodologies were developed to overcome the limitations of speed and scope of existing methods. Given the rapidity with which natural resources can be threatened in today's world, the biologist's multiyear, intensive study of a tiny enclave of "natural" vegetation is often as inappropriate for policy purposes as the anthropologist's multiyear study of a single, "traditional" community. The importance of these rapid appraisal methodologies lies, in part, in their ability to identify signs or surrogate parameters that can indicate the conditions of environmental health in a particular location, and the factors that both threaten and support it, in a relatively short time period.

A second contribution made by the authors in this volume pertains to the expansion of the spatial and temporal scales of analysis of questions pertaining to conservation and society. Regarding temporal scales, Puri demonstrates that the landscape observed today in Kalimantan is the consequence of generations of relations between the mobile Penan and their natural resource base. This deepened temporal perspective suggests that it is not enough for policymakers to understand resource use practices today because such snapshots in time do not tell us where biosocial processes are going and where they will be in the future. In order to design effective conservation policy, therefore, it is important to understand past as well as future trajectories of environmental change. Regarding spatial scales, Puri, Ticsay, Sajise et al., and Gunawan, Parikesit, and Abdoellah have all demonstrated that when agroforestry systems are seen not as islands of anthropogenic disturbances but as components of diversity at the larger, landscape level, their contribution to overall ecosystem health and species diversity can be better assessed. In their chapter on the Citarum river basin in Indonesia, Gunawan, Parikesit, and Abdoellah show that the failure of the state to consider landscape-level dynamics among farmers, their agricultural fields, and proximate forest resources has actually led to decreased stability and diversity at the regional level. These findings reflect an important thesis underlying much of the work in this volume: a critical determinant of conservation in many rural landscapes is the heterogeneity of the agricultural landscape, so that as this heterogeneity varies so, too, will conservation vary. These findings lend a new imperative to the re-thinking of work by agricultural scientists and extension workers who heretofore have been primarily concerned with maximizing crop productivity and economic benefits, typically through the promotion of monocropping technologies.

A third contribution of the book lies in providing a more nuanced view of the relationship between agriculture and conservation through in-depth local studies. Many of the chapters show that complex, composite agricultural systems can mimic, reflect, and promote natural biodiversity in ways that more "modern," intensive, market-oriented systems cannot. These findings complement recent approaches to the study of agriculture and agricultural change, which are paying more attention to the way that population/land pressure and increasing market orientation in semi-urban zones are changing, for the worse, the biodiversity of traditional gardens and farms (e.g., Mary and Michon 1994). Several of the chapters draw our attention to the deleterious effects of the Green Revolution on landscape and genetic diversity. This, in turn, has led us to ask, in what land use types (within a given agroecosystem) are most endangered species found, and what land use types themselves are most endangered? These questions reflect the emphasis of these chapters and the volume as a whole on the wider linkage between the environment and the social conditions required for its conservation. Many of the works in this volume address this linkage through their analysis of habitat fragmentation and landscape homogenization and conversion.

A fourth contribution of this volume lies in problematizing institutions and management, including institutions and management thought to be part of the solution to problems of conservation and development. Frossard, Lye, and Dove all demonstrate that many of the key institutions involved, including national parks, international research centers, and Western legal mechanisms, have done more to undermine than strengthen local practices that support conservation. These institutions all tend to foster local engagement with the global community, a relationship that is often problematic for the local social relationships surrounding resource conservation. Some scholars argue that *any* engagement with the global community is likely to be inimical to conservation. Thus Lohmann (1991) argues that in Thailand local people often have a vested interest in practices that are associated with resource conservation, whereas the business sector and national and international agencies often have a vested interest in just the reverse. He especially disagrees with the argument of many economists that the key to biodiversity conservation is greater engagement with the global market economy. He argues, indeed, that what is needed is more *dis*engagement (see Corry 1993b and Dove 1993b). Similar conclusions are reached by Dove, Frossard, and Lye in their analyses of intellectual property rights, rights to genetic diversity, and indigenous views of human environmental relations. Each of these analyses demonstrates that the local communities responsible for the stewardship of natural resources are often politically, economically, and culturally marginal and, as a result, it is unlikely that national elites will pass on to them the benefits being offered by the international conservation community. On the contrary, the fact that these marginal communities possess valuable biogenetic resources is likely to lead to greater, not lesser, disenfranchisement by covetous national elites. These analyses show that there is little agreement and much disagreement over who the victims, villains, and potential saviors are in the international battles over biodiversity conservation.

Finally, the contributors to this volume offer a new way to look at the transnational flow of concepts central to conservation discourses. The chapters by Dove, Lye, and Frossard all suggest that prevailing discourses are based on a narrow vision of how natural resources should be used and protected, a vision that emanates from Western scientific and policy institutions. Lye's analysis of a national park in Malaysia shows how conservation has become a powerful tool for strengthening state control over valuable natural resources by suppressing the social, economic, and political realities of indigenous peoples and by obscuring alternative ways of conceptualizing and using such resources. Likewise, Frossard and Dove both show that the discourses produced by international research institutions such as the International Rice Research Institute operate so as to enhance centralized control over valuable genetic resources at the expense of control by the people whose lives directly depend on them. All of these authors draw attention to the disjuncture between representations of scientific knowledge as absolute and the way in which such knowledge is actually received and transformed in particular times and places.

Taken as a whole, these analyses suggest that the solution to biodiversity conservation does not lie in the traditional model of separating people from nature. Recent efforts to "integrate" conservation and development (see Wells 1994; Wells and Brandon 1992; West and Brechin 1991; Western and Wright 1994) move us closer to a potential solution, but they still separate people from the protected landscape. A better solution lies in identifying local institutions that have the potential to conserve the environment through sufficient economic returns without long-term resource degradation. This may entail supporting agricultural practices that are not entirely market-oriented and fostering social institutions that contradict the Western logic of scientism.

In summary, this volume suggests that it is time to shift conservation efforts away from the externally driven introduction of new policies and programs at the local level and away from an exclusive focus on finding new ways to engage local communities with the global market. Instead it is time to more carefully listen to local voices, more fully understand complex local practices, and appreciate local categories and worldviews. It is also time to search critically for extralocal threats to conservation and development and to denaturalize the international institutions and scientific discourses that dominate work in this area. Through such attention to both local and extralocal realities, it may be possible to enhance and foster conditions that support a rich and varied landscape, a diversity of floral and faunal species, and equitable socioeconomic conditions. The key to achieving these goals is a radical rethinking and redrawing of the boundaries—not only spatial and temporal but political and conceptual—of the problems of conservation.

Part I:
Historical Perspectives on Conservation and Society

1

Postabandonment Ecology of Penan Forest Camps: Anthropological and Ethnobotanical Approaches to the History of a Rain-Forested Valley in East Kalimantan

RAJINDRA K. PURI

DURING LARGE MAST FRUITING EVENTS in Central Borneo's rain forests, which may occur every two to five years (Ashton, Givnish, and Appanah 1988; Caldecott 1988; Dove 1993d), semisedentary Penan Benalui leave their homes in the villages of neighboring swidden agriculturalists for temporary forest camps to gorge themselves on fruit, fat wild pigs, and processed sago palms (Puri 1997a). Fruit collected from the surrounding area is brought back to the camp where it is processed and shared among all residents. At the time of abandonment, two to eight weeks later, the camp is littered with piles of discarded seeds from more than a dozen tree species. Within a year, the site is covered with a carpet of seedlings. It is as yet unknown whether these actions alone will lead to the establishment of permanent fruit groves in the forest, but it appears that in some locations a preexisting *pulung bue'*, a grove of fruit trees, is at least perpetuated by this annual visit. Once established, the area may serve additional functions as a campsite for Penan and other collectors of forest products and, if close to current settlements, as a hunting patch. Its primary function remains, however, the production of fruit, a favored and important part of the diets of all peoples on the island of Borneo.

The Penan have several kinds of forest camps besides fruit camps. Some are fairly extensive and are designed for specific activities, such as processing sago palms (*Eugeissona utilis* and *Arenga undulatifolia*, among others), collecting nontimber forest products, or working in gardens and rice swiddens. These activities require extended stays, and so the camps may be occupied for several weeks at a time, periodically through the year. There are many more temporary forest camps that are used repeatedly but only for overnight or short stays while hunting, gathering, and traveling between villages or other camps. Except for very steep slopes and cold mountain peaks, forest camps are found almost anywhere. It is therefore not surprising that new and old sites are ubiquitous in their occurrence throughout the forests and mountains and alongside the streams and rivers of a Penan group's territory.

In this chapter, I describe the goals, methods, and some preliminary results of research directed toward understanding the long-term effects of these Penan forest camp sites and associated economic activities on the floristic structure and composition of lowland and mountain dipterocarp forest of the Lurah river valley, in East Kalimantan, Indonesia (fig. 1.1). This research involved studying the historical ecology of eighteen recent Penan forest camps (less than 40 years old) and two fruit orchards in the vicinity of a village of swidden farmers—the Kenyah Badeng of Long Peliran at the mouth of the Lurah River. Data collection methods were drawn from ethnohistory, ethnobiology, and forest ecology, while the analysis relied upon multivariate statistical techniques to compare the vegetation structure and floristic diversity of the camps and discover the potential underlying social and ecological dimensions responsible for the observed variability.

The data collected during the research is here analyzed with regard to the following questions. How are forest camps chosen and what activities are associated with establishing and using them? How do we identify and date these camps once they are abandoned? Do Penan activities associated with their campsites change the structure and floristic composition of the surrounding forest and for how long? Does this impact increase, sustain, or degrade plant species' abundance and diversity? What biophysical or social factors

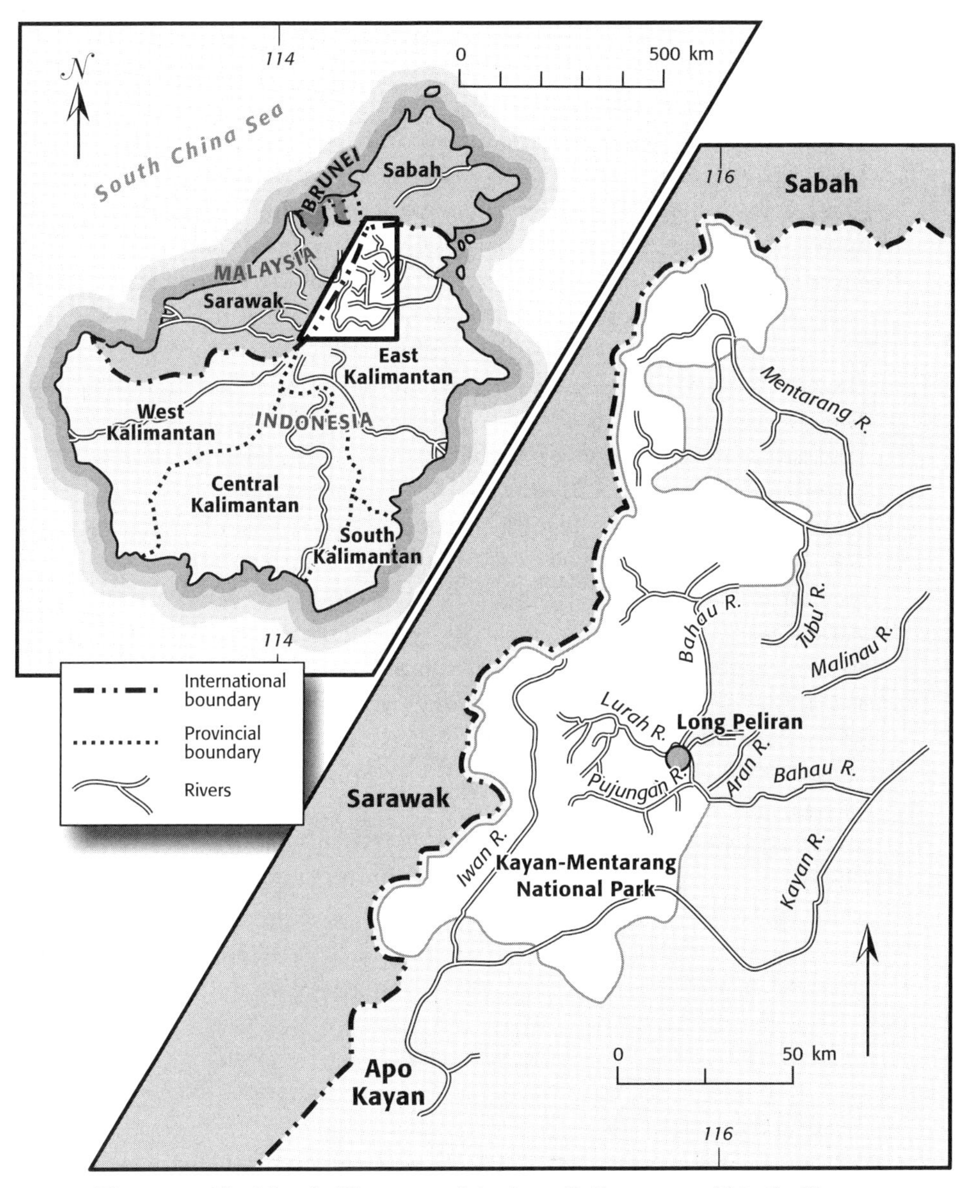

Figure 1.1 The island of Borneo and the Long Peliran area within the Kayan Mentarang National Park

are characteristic of sites and related Penan activities that show high floristic diversity?

The answers to these questions provide insight into the place of the Penan in the history of these forests. If the Penan play an important role in creating the observed floristic composition, and by extension influence animal behavior patterns, then that will have important implications for the formation of forest management policy and economic development proposals for the Lurah Valley and surrounding areas now included in the Kayan Mentarang National Park. More generally, the research addresses the role of humans in the creation and maintenance of vegetation communities in a rain forest.[1] While research and intense political debate on the loss of biodiversity is well under way, research on the factors that promote and sustain biodiversity is still rather maverick,[2] analogous to studying people who do not get AIDS rather than studying HIV and its victims. In both cases, researchers are searching for the underlying conditions that result in health. Unfortunately, success stories in either case are exceedingly rare. While some research has demonstrated that tropical forest farmers, past and present, coexist with high levels of biodiversity, and in some cases may even be responsible for those levels of biodiversity (see Meilleur 1994), this study addresses the consequences of the activities of nonagricultural societies (see Roosevelt et al. 1996). The typical view of forest hunter-gatherers as benign denizens living "with" the rain forest is reexamined in light of the study's premise that there may be long-term cumulative effects of purposeful, though sometimes subtle, human manipulations of vegetation. These issues are discussed following the presentation of the research results.

Background

The Lurah River Valley

The study area is the Lurah River, located in the district of Pujungan, in northwest East Kalimantan, which flows east for 65 km from the mountains bordering the Iwan River valley to its mouth at the Bahau River (see fig. 1.2). The valley is roughly 2,000 sq km in area. At the river's edge, the altitude is between 250 and 600 m above sea level

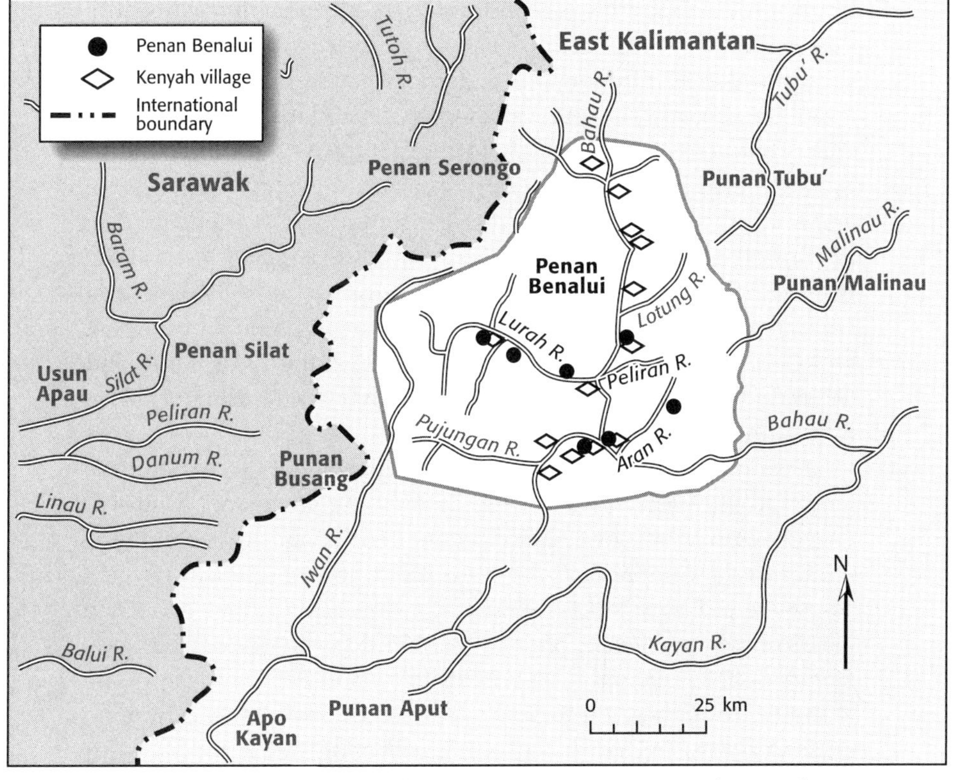

Figure 1.2 Penan Benalui settlements in 1998 and estimated range during the twentieth century. The map does not show settlements of relatives resident in the Malinau river valley and the Usun Apau area of Sarawak.

(m asl) and rises to 1,600 to 2,500 m asl at mountain peaks to the north and south. This part of Borneo is mountainous, with fast flowing streams, lowland dipterocarp rain forest to roughly 500 m asl, oak-laurel forests to 1,000 m asl, and montane forests on the higher peaks. The climate is typical of the equatorial Tropics: hot, humid, and wet. The driest months are usually June through August, while the wettest months are September through January (Puri 1997a).

The known history of the Lurah river valley, as reconstructed through studies of oral history and genealogy (Sellato 1994, 1995, 2001), begins in the early seventeeth century; its prehistory is as yet unstudied. Basalt adzes, stone burial jars, and dolmens attributed to the Ngorek peoples have been found in settlements that probably date to the seventeenth century or earlier. Since then, there have been several waves of migration through the valley from the neighboring Pujungan, Iwan, and Bahau rivers, with most originating in Sarawak. Thus, the valley contains former village sites of the Ngorek; Kayan; mixed Kenyah-Ngorek groups such as the Nyibung, Merap, and Pua'; and various isolect (linguistic) groups of the Kenyah (Puri 1997a; Sellato 1995, 2001; Vom 1993). Since the Penan arrived at the end of the last century, some Kenyah groups, Uma' Badeng and Uma' Bakung, have occupied villages at Long Peliran, Long Apan, and Long Bena, while other Kenyah groups, the Uma' Long and Leppo' Ndang, have only stayed briefly while migrating from the Pujungan River to the Bahau River.

This long history of human settlement, migration, and subsistence activities has resulted in the creation of a patchwork of sites of varying age across the valley. To date, more than fifty abandoned villages and campsites have been identified (Puri 1996, 1997a). There are old village sites either on inaccessible ridges or on the banks of the main rivers, often identifiable by orchards of fruit trees, stone burial jars, or old fortifications. New and old forest camps are scattered throughout the area and are used as shelter during hunting or sago-processing trips and while collecting forest products for trade or fruit during a mast fruiting event. Old gardens and rice fields are also found, many in very late stages of succession but still identifiable by local informants as being secondary forest. There are fields dominated by cogon grasses *(Imperata cylindrica)* in areas said to be old rice fields, which are still regularly burned by local people to attract deer. There are also numerous trails that are maintained and used for traveling between villages, hunting, forest product collecting, and migrations between neighboring watersheds. All of these anthropogenic environments are used today, especially the abandoned village gardens, which are favored for fruit-collecting and hunting grounds.

Today, the Lurah river valley and surrounding Pujungan District has a population of more than three thousand people in twenty-one villages, representing eleven indigenous ethnic groups, including Kenyah, Penan, Pua', and Saben people. They practice agriculture and use forest resources in a wide variety of ways, including the cultivation of tuber gardens, mixed crop swiddens, irrigated rice fields, and small plantations of such cash crops as banana, coffee, cacao, and cinnamon. They manage and harvest forest-grown fruit trees, sago palms, rattan, and aromatic woods for trade. Food, materials, and trade items are also obtained by means of fishing, hunting, trapping, and gathering. This diversity is found not only on a regional scale but even within single communities, where people often combine or switch between different forms of resource use (Puri 1997a).

The Penan Benalui

The Penan Benalui of the Lurah River are an offshoot of the Western Penan of the interior of Sarawak, Malaysian Borneo (Needham 1972). A group of probably 50 to 75 migrated to the Lurah River Valley from Sarawak about one hundred years ago under the protection of their patrons, the Kenyah Badeng, who are longhouse-dwelling swidden agriculturalists. In 100 years, these Penan have ranged across an area of nearly 10,000 sq km, making countless forest camps and expanding to more than 350 people in three official villages and three communities within Kenyah villages. The Penan began settling down, a process of sedentism and increased dependence on agriculture, in the mid-1960s. There are currently four Penan settlements on the Lurah River, Long Peliran (the primary study site at the mouth of the Lurah), Long Belaka (on the middle Lurah), and Long Sungai Ma'ut and Long Bena (on the mid-upper Lurah). Long Peliran and Long Bena are also Kenyah Badeng villages (Puri 1997a).

The traditional economy of nomadic Penan Benalui consists of hunting (primarily the bearded pig, *Sus barbatus*, and other terrestrial vertebrates), collecting wild plant foods (of which wild forest fruits are most important) and animal foods (such as freshwater sago grubs, snails, turtles, and frogs), processing palms for their starch-filled pith (sago flour, from *Eugeissona utilis* and *Arenga undulatifolia*), and collecting and trading forest products such as rhinoceros horn, bezoar stones (from *Presbytis hosei*), eaglewood or *gaharu* (*Aquilaria* spp.), and rattan canes (predominantly *Calamus caesis*). Penan trade forest products for salt, sugar, tobacco, cloth, metal, dogs, rice, and garden vegetables. Fishing, trapping, and cultivating manioc *(Manihot esculenta)* and rain-fed hill rice are minor economic activities, although for some families their importance in the diet has grown. These activities are commonly conducted in and around the forest camps that are the subject of this study.

Almost all foods are immediately shared among all families in a group, although hunters usually reserve choice parts for their families (e.g., fat layers and internal organs) and the owners of sago palm clumps and fruit trees claim larger shares (Puri 1997b). Cultivated plant foods such as rice and manioc are not immediately shared among families but will be given if requested. Of importance to this study is the fact, already mentioned, that Penan fruit collectors bring harvested fruit back to their forest camps to be consumed by all members of the group. This socially accepted practice has the unintentional effect of clumping a variety of species of discarded fruit seeds in a cleared area.

Nomadic Penan groups range in size from 20 to 100 people. Today's semipermanent communities, some located in the Kenyah villages, vary in size from 20 to 130 people. A Penan family, which could be nuclear or extended, is recognized as composed of those who share a cooking hearth. An elevated forest shelter, or *lamin*, usually houses just one family, ranging in size from 2 to about 10 people. In the past, forest camps of several *lamin* were occupied from one to three months, then temporarily abandoned. After several years, the site might be used again as a main settlement or in the meantime short-term camps might be established there for particular purposes. Historically, nomadic Penan primarily chose settlements based on the presence of sago palm stands (usually *Eugeissona utilis)* (Brosius 1991; Puri 1997b), but they also had favored places for fruit season camps. Camps were usually located in clearings in secondary forests, for example, in swidden fallow or old tree-fall gaps in mature forest, because of a justified fear of falling trees and branches. Penan prefer to locate their settlements on flat banks next to streams but will establish camps on flat ridgetops or mountain plateaus if water is available nearby.

Campsites are abandoned because of disputes, because food resources are scarce (e.g., sago palms, wild pigs, and fruit), or because disease or death strike. Sudden death, for instance, from a falling tree, is never seen as accidental but instead as an omen of bad relations with the spirit world. The ghost of the departed is said to haunt the area, and thus the Penan move far away and avoid that site for a long time. It will probably never be knowingly used again by the same group. For the same reason, Penan are wary about establishing camps in old abandoned villages or forest campsites, because they do not know whether people died there or if magical artifacts that may make them sick are buried in the area. Because ghosts, the effects of magical artifacts, and disease tend to travel downriver, both Kenyah farmers and the Penan prefer to build new villages or camps across or just upriver from older ones. Over several generations, this settlement behavior results in constant movement and the creation of a large number of forest camps across the landscape.

Penan society is basically nonstratified and egalitarian, with some authority to settle disputes residing in a respected elder, who is usually the patriarch of all the families in the group. The Penan rely on their leaders to deal with the outside world, meaning that they must be knowledgeable about other languages and customs as well as skillful negotiators. Despite this role, few leaders have complete control over individual families, which regularly, and freely, pick up and leave because of disputes. The death of a patriarch can lead to fission of the group because brothers and sons of the deceased will compete for power. Marriages between the children of feuding factions and threats from warring neighbors are reasons for these groups to reunite. Splits in Penan communities have occurred many times in the hundred years they have lived in the Pujungan District, hence the total area covered by all subgroups amounts to more than 10,000 sq km, with literally hundreds of forest camp sites established in that time.

As important as the dynamics of internal social relations, external relations with farmers and traders have also been factors in determining movement and settlement patterns. When still predominantly nomadic, the boundaries of a Penan group's range and the locations of its campsites were often determined by the farming groups with which these groups traded regularly. Most of these trading relationships were of the classic patron-client type, in which the aristocratic leaders of the stratified farming societies, such as the Kenyah Badeng, had exclusive trade agreements with the leaders of the Penan group. Despite an increasingly sedentary lifestyle, Penan still maintain similar economic relationships with farmers and wealthier neighbors.

The Penan of the Lurah have always maintained alliances with the most powerful of Kenyah leaders in the area, usually settling within a day's walk of the village (see fig. 1.2). In the past, many Penan men hunted and collected valuable animal and plant products for their patrons, who then sold them to traders and merchants downriver. Disputes between Penan and their patrons, usually over debts or claims of deceit, sometimes caused whole villages of Penan to move in search of new patrons. Each of the Penan groups has had a different historical relationship with its neighbors, resulting in villages with economic, political, and social ties often varying at the level of the individual household.

Today, the Penan of Long Peliran and Long Bena are still somewhat restricted in their use of local resources because the Kenyah maintain control over major land use decisions such as the location of new rice swiddens. The Penan in Long Belaka and Long Sungai Ma'ut live in their own villages and are thus more independent, but certainly they are not free from the influence of outsiders. Typical of other remote groups of people in Indonesia, the Penan are under great pressure to settle down, adopt agriculture, attend schools and church, and participate in the development of an Indonesian national culture. Most of the lands of the Penan are now included in the Kayan Mentarang National Park. There have been, and will continue to be, opportunities for the Penan to participate in the research and development of a management plan for the park and local economic development activities in the buffer zones surrounding it. In general, as the process of sedentism continues, Penan families spend more and more time in and around permanent villages, establishing forest camps closer to the village proper—often in highly impacted forests—and going farther than a day's journey when on expeditions to collect forest products.

Methods

In order to discover the impact of Penan settlements on the forests of the Lurah, field research was conducted in Long Peliran in 1996. It involved three main tasks: (1) locating and dating forest camps through interviewing Penan and others about their migration history, (2) participating in forest expeditions to study how forest camps are established and their immediate effects on the forest, and (3) establishing sample plots in forest camps and comparing stand structure and floristic composition. Additional information about the local ecology and economy was collected by the author during 1991–1993 (Puri 1997a, 1997b). Comparative information about the ecology and distribution of fruit species were derived from studies in the Apo Kayan and other parts of East Kalimantan (Valkenburg 1997). Although preliminary, inventories conducted by Project Kayan Mentarang of the World Wide Fund for Nature–Indonesia Programme (WWF-IP), and the Center for International Forestry Research (CIFOR) in the neighboring Bulungan Research Forest, provide some indication of the floristic composition of undisturbed forests in the study area (Blower, Wirawan, and Watling 1981; McDonald et al. 1992; O'Brien 1998; Wulfraat and Samsu 2000; Sheil et al. 2002).

Ethnohistory and Ethnoecology Interviews

The initial phase of the research involved interviews with local informants, individually and in groups, concerning the location and character of past and present Penan habitation sites in the Lurah river valley (see appendix, table 1.1A). Both current residents and those who had moved to new villages were interviewed. For all locations, detailed information was collected concerning the age of the site, the duration of Penan occupation, the activities carried out there, and the reasons for abandonment. Informants were also asked to identify indicators of past habitation and associated activities. Finally, older informants were asked to identify, and when possible date, trees that they had planted in these forest camps as well as those in village gardens and swiddens. This information is critical for understanding the process of replacement of secondary forest and garden species with mature forest species.

Spatial Data

A handheld Global Positioning System (GPS) unit was used to provide location data of the established plots (see fig. 1.3.). A Landsat™ image of the area, provided to Project Kayan Mentarang by the East-West Center's Spatial Information Lab, was used as a base map for this study. Location data generated by the GPS were used to calculate the distance from campsites to the village of Long Peliran. The campsites are too small to be accurately located or even seen on the Landsat image; thus, no analyses were conducted to determine whether there might be a unique spectral signature for vegetation associated with these camps.

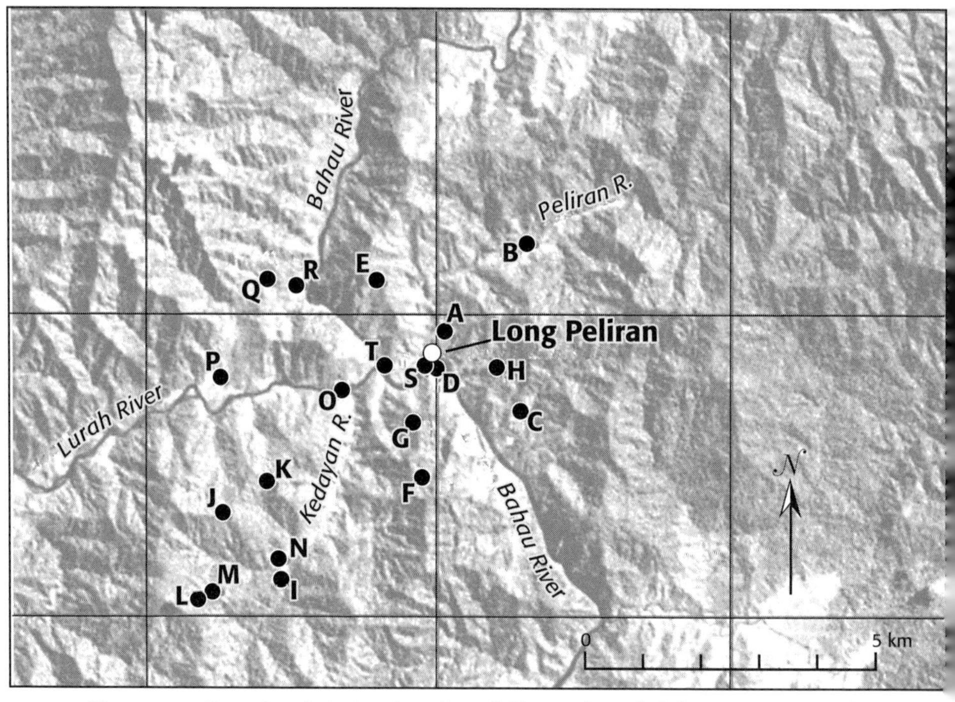

Figure 1.3 Sample plots in abandoned Penan Benalui forest camps and Kenyah Badeng orchards in Long Peliran area. Plots D, O, and S are in orchards. Data from Landsat™, June 1990, color composite bands 3, 4, and 5. Courtesy of the Spatial Information Lab Program on Environment, East-West Center.

Focal Expeditions

Information concerning present day use of forest areas was collected during the field research.[3] Through participant observation, information was collected on the current use of forest camps and surrounding vegetation for collection of fruit, *gaharu*, rattan, and other forest resources used for subsistence and trade. Data were collected on the size of cleared campsites, the identity and abundance of cut and coppicing trees, and the abundance of cultivated plant and edible fruit species.

Analysis of Vegetation Changes in Forest Camps

Eighteen Penan forest camps and two fruit orchards were chosen for vegetation studies and characterized according to several anthropological and biophysical variables (table 1.1; for data, see appendix tables 1.2A and 1.3A).

Each site was sampled with one randomly placed 40x10 m plot (0.04 ha). These transects were long enough to transit the apparent boundaries of former shelters and work sites and sometimes included surrounding vegetation. It is probable that human impacts on the vegetation extend much further than these plots could capture. Floristic diversity in sample plots was studied initially using Penan taxa instead of botanical species, as many plant species are as yet unidentifiable. Penan categories of ecological representation were used to characterize the vegetation of the sites (Puri 1997a). All trees were labeled with a numbered aluminum tag so that they could be used repeatedly over the course of the study and possibly in the future by staff of the Kayan Mentarang National Park. Herbarium voucher specimens were collected from all inventoried Penan taxa by means of tree climbing. The specimens were identified with the assistance of botanists at the Wanariset Herbarium, Balikpapan; the Rijksherbarium/Hortus Botanicus, Leiden; and the Herbarium Bogoriense, Bogor, Java. A set of specimens was also sent to Project Kayan Mentarang's Lalut Birai Forest Station. Following methods described by Peters (1996) and Valkenburg (1997), each plot was sampled by recording the following information for all trees over 10 cm in diameter: vernacular name, local use, location on a site map, diameter at breast height (dbh), and height (to first fork and total

Table 1.1 *Anthropological and biophysical variables of sample plots*

Variable	Code	Explanation
Primary activity at site	ACT	Coded: 1 = residence, 2 = fruit camp, 3 = swidden camp, 4 = sago camp, 5 = NTFP expedition camp
Population size	POP	Estimated number of families (1–6)
Number of human occupations	OCC	Estimated number of times Penan families occupied site
Duration of occupations (yrs)	TIM	Estimated length of time of residence at forest camp
Time since abandonment (yrs)	ABN	Estimated length of time since last occupation
Vegetation type (from most to least manipulated)	VEG	Coded: 1 = village gardens, 2 = young secondary, 3 = secondary, 4 = old secondary, 5 = mature forest, 6 = primary forest
Forest age (yrs)	AGE	Estimated from interview data
Altitude (m asl)	ALT	Measured using handheld altimeter
Distance to Long Peliran (m)	DIS	Measured on direct line between site and plot S in Long Peliran
Site location and human disturbance (from most to least disturbed)	TOP	Coded: 1 = riverine, 2 = riverine with trail, 3 = streamside, 4 = streamside with trail, 5 = hillside terrace, 6 = terrace with trail, 7 = ridgetop, 8 = ridgetop with trail

height). Smaller plots of 5x5 m (6.25 percent of the plot area) were similarly sampled for saplings over 1 m in height. Plots of 1x1 m (0.25 percent of the plot area) were sampled for seedlings less than 1 m in height, although these data are not analyzed here.

Using this information, the sample plots were compared on the basis of several stand structure variables, diversity measures, and floristic compositions. Differences in the structure of vegetation were expected to reflect disturbances associated with clearing and burning sites. Variability in the diversity and abundance of taxa was expected to show the effects of natural processes of succession in the

Table 1.2 *Variables describing stand structure of sample plots*

Character state	Code	Explanation
Tree density (trees ha^{-1})	DEN	Number of trees per plot area
Basal area (m^2 ha^{-1})	BAS	Total basal area of trees per plot area
Diameter (cm)	DBH	Average diameter of all trees > 10 cm dbh
Height (m)	HGT	Average height of all trees
Furcation index	FUR	Average of all trees: height to first fork/total height

sites as well as evidence of human manipulation of that vegetation through selective cutting and planting of preferred taxa. Stand structure was described by five character states (table 1.2).

The stand structure of the sample plots was compared using cluster analysis and principal component analysis (PCA). The PCA output produces a three-dimensional plot displaying the clustering, or similarity, of the forest camps. Analysis of the factor loadings provides an indication as to which variables underlie or explain the similarities and differences among the forest camps. A similar comparision, using principal coordinate analysis (PCO), was conducted using the above stand structure variables and additional quantitative and qualitative characteristics, both ecological and anthropological. As with the PCA, examination of factor loadings for these variables indicates which are correlated and which serve to explain the patterned variation observed in the clustering of sample plots.

Floristic composition and diversity were described and analyzed in several ways. First, taxonomic richness, a measure that relates the number of taxa to the number of individuals in the plots, was compared using Margalef's index (MGT, MGS), and Fischer's index (FST, FSS) (see Magurran 1988). It was expected that richness would be lowest in fruit gardens and orchards and highest in the oldest abandoned camps in primary forest. Low richness in the latter sites might indicate long term effects of human disturbances such as clearing, selective felling and planting of preferred taxa (table 1.3).

Table 1.3 *Diversity variables of sample plots*

Variable	Code	Explanation
Margalef's index trees	MGT	Measure of richness (relates number of taxa to individuals)
Margalef's index saplings	MGS	Measure of richness (relates number of taxa to individuals)
Fischer's index trees	FST	Measure of richness independent of plot size
Fischer's index saplings	FSS	Measure of richness independent of plot size

Second, taxonomic dominance (DOM), that is, which taxa were most abundant in the plots, was determined by comparing the total basal area of all individuals for each Penan taxon. Dominance was expected to be indirectly related to taxonomic richness, such that high dominance by a single taxon should be found in plots with low taxonomic richness. Dominance of edible fruit species was expected in fruit gardens and fruit camps, while dominance of *Vitex pinnata* (VDO) was expected in camps established in old swiddens. Dominance was also calculated for groups of taxa, such as edible fruit taxa, and indicators of human disturbances, including the dominance of *Vitex pinnata* and the dominance of coppiced trees. It was expected that fruit diversity (FRT) and fruit dominance (FDO) would be high in abandoned fruit camps that were used frequently. *Vitex pinnata* was expected to dominate sites made in former swiddens. Resprouting of felled trees, or coppicing (CDO), was expected to be greater in a campsite where an area for a shelter had been cleared. Since most coppiced trees bifurcate at the stump, they also have a low furcation index (FUR). Thus dominance of coppicing was expected to be indirectly related to the average furcation index. These measures serve to highlight different kinds of human disturbances associated with establishing and living in forest camps (table 1.4).

A third way to measure floristic diversity in the campsites was to compare the taxonomic abundance of the plots. This diversity measure, which accounts for the variable abundance of taxa as well as

Table 1.4 *Dominance measures of sample plots*

Variable	Code	Explanation
Edible fruit taxa (% of all trees)	FRT	May indicate the presence of a fruit camp or village garden
Dominance of fruit trees (% of basal area)	FDO	May indicate the presence of a fruit camp or village garden
Dominance of *Vitex pinnata* (% of basal area)	VDO	May indicate extent of disturbance for agriculture
Dominance of coppiced trees (% of basal area)	CDO	May indicate extent of clearing for campsite
Taxonomic dominance (%)	DOM	Taxa with highest proportionate basal area; may indicate human influence

simple taxonomic richness, was analyzed by first calculating the abundance for each Penan taxon inventoried in the sample plots. Abundance was calculated by summing the biomass (estimated by multiplying basal area by height) of all individuals. These values of taxonomic abundance were compared among the plots using a correspondence analysis program in the Numeric Taxonomy System (NTSYS), a statistical software package. As with PCA and PCO analysis, the procedure produces a scatterplot showing clustering of the sample plots and factor loadings showing which taxa were important in determining the clustering of sample plots. It was expected that taxonomic abundance would be highly reflective of vegetation type and thus the type and degree of human manipulation that might have occurred at the site. The analysis would show how the structure of the vegetation, especially the volume of trees, was influenced by the manipulations required to physically construct the campsite. It would also show how the richness of the vegetation was influenced by decisions to foster or promote certain categories or even particular species of trees and exclude others. Therefore, it was expected that the clustering of sample plots based on taxonomic abundance would be similar to, and thus would reinforce, the results of the other diversity tests and the scatterplot of the PCA analysis of stand structure variables.

An important methodological problem that arose during this study was dating campsites of unknown age. It was thought that the size of the tree vegetation would be a good indicator of age since abandonment, so a number of variables were tested that might reflect tree growth in order to date the standing vegetation (table 1.5). First, the average girth (DBH) and height (HGT) for all trees in a sample plot was tested against the estimated ages of the forest (AGE) and abandonment (ABN).

The underlying assumption, that all trees were in fact affected by human habitation proved to be false because many trees are left untouched during occupation of a camp. The issue of dating is further complicated by the fact that the affected standing vegetation may derive from several settlements occurring at different times. The age of the campsite may be much older than the last occupation or the largest and oldest standing tree or it may be much younger. Thus, additional indicators of growth for specific groups of trees and particular species were also tested. The size of the largest tree (BIG) in each plot was expected to give an indication of the upper limit on the age of the forest both in and around the campsite. The size of fruit trees (FRD) was expected to be a good indicator of age since abandonment, especially if the species are commonly planted and cultivated. As already mentioned, the size of *Vitex pinnata* individuals may serve as an indicator of the length of time since the last burning. The size of coppiced individuals, measured by average diameter

Table 1.5 *Tree growth variables tested against age of site vegetation*

Variable	Code	Explanation
Largest diameter tree in sample plot (cm)	BIG	May indicate the age of the forest on or around site
Average dbh of all edible fruit taxa (cm)	FRD	May indicate the age of the forest on site as well as age since abandonment
Average dbh of *Vitex pinnata* (cm)	VIT	May indicate length of time since burning for swiddens
Average dbh of coppiced trees (cm)	COP	May indicate length of time since clearing

Note: cm = centimeter; dbh = diameter at breast height.

(COP), may serve as an indicator of the length of time since last clearing and therefore should be directly related to the age of the site.

Results and Analysis

History of the Penan in Long Peliran

In the 1950s, five families of Penan moved from the Aran river valley to the village of Long Peliran, a two-day journey on foot (fig. 1.1). Several elders claimed that they were drawn to the area by promises from the Kenyah Uma' Badeng leaders that relations would be friendly and beneficial for both groups; the Kenyah often sponsored forest product collecting expeditions and liked to have Penan along as guides and hunters. In order to persuade them, the Badeng reminded the Penan of the close relationship that once existed between their ancestors, even before their common migration from Sarawak in the 1890s. The Penan of the upper Aran River had broken their ties with the Badeng in the 1920s and then moved away from their territory in the upper Lurah river valley. They eventually became allies of the powerful Kenyah Uma' Alim in Long Metap, a village near the confluence of the Aran and Bahau rivers. By the 1950s, these Penan had begun to feel unwelcome pressure from the Uma' Alim to settle down, adopt farming, and become Christians. Thus, the migration of five Penan families back to the Kenyah Badeng turned out to be mutually agreeable to both sides. Many Penan families remained in the Aran river valley, however, eventually establishing a village at Long Lameh in the 1960s (Puri 1997a).

The reconstructed history indicates frequent seasonal and annual movements of the Penan within the village territory of Long Peliran between the late 1950s and 1996. At least fifty-one moves among thirty-one campsites are documented in the Penan oral history for just one of these families (see appendix, table 1.1A). The total number of former Penan campsites within the Long Peliran territory is estimated to be between fifty and seventy-five. The number of Penan families occupying these camps has fluctuated due to the varying functions of the sites and the numerous fission and fusion events that characterize Penan family histories. In general, though, there has been a decrease in the Long Peliran Penan population due to out-migration. The same can be said for the Kenyah Badeng farmers, who have steadily moved downriver since the early 1980s. As of 1998, all the Penan had left Long Peliran for neighboring villages. The Penan elders that led the original migration are now dead, save one elderly woman.

Over the past generation, the Penan have moved closer to the Kenyah farmers in Long Peliran. The analysis below shows that the older camps were located farther from the village, at a higher altitude, and mostly in primary forests. One consequence of this changing settlement pattern is that older campsites are infrequently reused and thus are more likely to have a floristic composition typical of a site undergoing natural regeneration. In contrast, many of the recent Penan camps have been established in Kenyah swidden gardens, fruit orchards, and village home gardens. There has also been an increase in the number of camps in and around Penan swiddens, demonstrating a change in subsistence practices as well as settlement patterns in the last generation. The camps near the village are used more frequently than those in the upland areas and thus are subject to continuous manipulation of the vegetation by people in order to maintain access to and exploit their valued resources.[4]

Description of Forest Camps

Larger camps, usually in villages or swidden fields, range in size from 400 to 1,000 sq m, not including surrounding gardens, while smaller forest camps range in area from 100 to 400 sq m. The environmental effects of establishing such seemingly small sites, however, extend beyond the immediate shelter and living space and well into the surrounding forest. Activities associated with gathering materials, constructing the camp, and provisioning the inhabitants all effect the surrounding vegetation. A typical small forest camp to be used by half a dozen forest product collectors, for example, is found perched on a cleared riverbank overlooking a fast-running stream, but if circumstances demand it the Penan can build a shelter anywhere, even in a tree. The size and complexity of the shelter and kitchen, the most prominent features of a camp, are dependent upon the time available to build a camp, the expected length of occupation, and the creative energies of the inhabitants. A simple overnight shelter is not very elaborate, often a plastic sheet hung over a crossbar with bedding material of cut leaves. In times of impending darkness and fearful of snakes, ghosts, and sickness, the Penan often just sleep next to a fire on their rattan or pandanus leaf mats. A successful hunt may require a more elaborate shelter with a kitchen for drying, smoking, and cooking meat. Expected long-term campsites, established in areas of abundant game, fruit, or other forest products, may take several days to build and be quite extensive. Industrious children are often responsible for sprawling and elaborate campsites with their own private shelters and play areas.

A site is chosen carefully to avoid potential hazards such as standing dead trees, overhanging dead branches, or freestanding large trees (which can be blown over in strong winds). Thus, camps in short secondary forest are preferred. Dangerous trees may be felled if the site is otherwise a good one, but most large trees are usually left standing. They may serve as corner posts for a raised sleeping platform; their bark may be stripped for flooring, roofing, or cooking material; or they may be covered with the scars and graffiti of exuberant, machete-wielding children. Saplings are felled to open up the canopy to sunlight and then are used in constructing a shelter. While overgrown brush, rattans, and thickets surrounding a site are generally welcome as a protective measure against wild animals, the undergrowth on a site is usually cut back, especially if the occupants are going to sleep directly on the ground instead of constructing a platform. Removing the undergrowth also reduces the abundance of leeches and insects and avoids chance encounters with snakes. Leaves from fan palms, wild ginger, and other herbs are gathered for bedding or roofing material. The Penan clean house, so to speak, by cutting and splitting dead wood for cooking fires, hacking trails to the water's edge or a nearby ridge trail, and constructing a cooking hearth with accessories for drying and cooking meat and storing food, water, and kitchen gear. River rocks may be brought up to the site for seats, for elevating cooking pots that cannot be hung above the fire, and for sharpening knives, machetes, and spears. After a day of such activity, a previously shady, leaf-covered, wet forest floor is transformed into an exposed, hard-packed dirt floor.

Most campsites used for hunting, sago processing, or forest product collecting have specific areas for processing the harvested products. Rattan canes are cut into transportable lengths (2 to 5 m) and then dried on racks or clotheslines in a bright, hot clearing, or on the hot rocks of a riverbank. *Gaharu* collectors often construct platforms where they sit and whittle away the waste wood from around the valuable resin-impregnated pieces. Hunters build huge cooking fires and high platforms to dry or smoke large amounts of meat. When rendering the fat layer of wild pigs, they establish large cooking fires, bark vessels are made to hold the cooling fat, and bamboo sections from *leupek tup (Dendrocalamus asper)* are gathered to serve as storage containers. Harvested sections of sago-producing tree palms may be brought to a cleared area next to the campsite, where they will be split, the pith extracted, and then made into edible starch flour (Puri 1997b). Processing and consumption of fruit result in piles of discarded rinds and seeds at the forest edges of the site; these waste heaps can reach a meter in height by the end of a long fruit season. Unfortunately, most of these activities leave no identifying permanent mark on a campsite once it has been abandoned; degradation of the predominantly vegetative material is as fast as the regrowth that quickly overwhelms the site.

Identifying Abandoned Camps

The young Penan research assistants working on this field study visited many of these camps for the first time. They were as surprised as the author to be told by their elders that they were "in" an old camp. But the telltale signs were all around them, and within minutes they were turning up evidence of the past. A clump of fruit trees, all of which are favored edible species, on fairly level ground, near a water source, or alongside a trail are all good clues of past human settlement. Often the best evidence is in fact the current occupation of a site, as most good locations are used over and over again. As mentioned earlier, newer sites are often located next to or across the river from older ones, taking advantage of a choice location (topographically as well as ecologically) yet not risking the dangers of living atop the remains of one's ancestors. Evidence of older occupations is most noticeable in the vegetation, for instance, in the diversity and size of fruit trees. The Penan identify *pangin (Mangifera pajang), lasiu (Lansium domesticum), keramo (Dacryodes rostrata), paken (Durio zibethinus), kian (Artocarpus odoratissimus), meu (Nephelium ramboutan-ake),* and *jilen (Dimocarpus longan)* as excellent indicators of past human presence, especially if they are found in close proximity to each other. Other indications of former human habitation can be seen in the vegetation structure, such as the low branching or numerous forks resulting from frequent pruning, a clearing among the trees large enough for a shelter, or evidence of coppicing from trees that were cut where the shelter probably stood. A pile of easily transported rocks where they really shouldn't be is taken as good evidence of human occupation, especially on ridge tops far above stream beds. At some sites, trees with scars from being marked long ago are evidence of humans, but because this is common practice when traveling in the forest these scars alone cannot be taken as evidence of a campsite.

There are also difficulties in distinguishing Penan sites from the gardens and orchards established by the Kenyah farmers at their swiddens, especially because many of the more recent Penan campsites have been established in these old swiddens. Older campsites surrounded by secondary forest characteristic of swidden fallow presumably contained Kenyah swidden shelters, especially if they appear to date from the time when the Penan were still nomadic and camped in the headwaters regions. Without information from living descendants, it is very difficult to determine the identity of former occupants.

Dating Abandoned Camps

Dating Penan campsites without detailed archaeological study is also a very difficult proposition. For this study, descendants' oral histories and reconstructed genealogies (e.g., birth locations of living members) have been used to estimate the dates of occupancy (see appendix, tables 1.1A and 1.2A). As far as is known, the oldest sites examined in this study are the two Kenyah fruit orchards (sites D and O), both of which were established next to longhouses built in the 1940s. Ancestors of the Penan reportedly occupied one old campsite (L) in the upper Kedayan River valley during the early 1930s, after the Kenyah Badeng had moved to Long Peliran. At that time, most Penan families were living on the upper Lurah and only occasionally visiting and trading with the Kenyah in their new village at Long Peliran.[5] Precise dates for prior occupations of site L, and all other sites with similar histories, are currently unavailable. Carbon-14 dating of the remains of hearths would probably be inaccurate, as the method is too coarse to date such young sites. Potsherds, found scattered about abandoned Kenyah villages, could be dated, but none has been found in Penan campsites. Even if they were, their provenance would be in doubt as there is no pottery tradition among the Penan, who instead used bamboo, bark, and gourds or traded for cooking pots. One method under consideration is the use of a form of tropical dendrochronology that might be applied to known planted trees (see Bormann and Berlyn 1981).[6]

Another technique would be to estimate the relative age of sites by using tree growth data for species found on the site. Because large trees may never be affected by the disturbances characteristic of human occupation, there is no necessary correlation between the age of the forest (or the trees on the site) and the age since abandonment. Conditions necessary for this test to work, however, might include trees known to have been planted while people were living there or trees known to have sprouted or coppiced in formerly cleared areas, for example, a space cleared for a sleeping shelter. Even with this knowledge, dating sites using rates of tree growth is complicated: trees grow at different rates, and growth may be affected by a variety of site-specific biophysical characteristics, from soil quality to rainfall to protection from violent winds (see Ashton 1981). Nevertheless, several hopeful candidates for indicator species were tested against estimated ages of the standing trees at the site or in the surrounding vegetation (AGE). In table 1.6, the diameter of domesticated and semidomesticated fruit trees, the common indicator of swidden fallow *Vitex pinnata*, and all coppiced trees in the examined campsites mostly proved to be positively correlated with the estimated age of the vegetation, albeit at levels of low significance. When data from all plots were available, the average diameter of all trees (dbh>10 cm) proved to be the best predictor of the age of forest vegetation,

Table 1.6 *Results of regression analysis of tree growth variables on estimated age of forest vegetation*

Independent Variable	Explanation	Sample Plots	*R*	R^2	R^a	R^{2a}
DBH	Average diameter of all trees	20	0.73	0.54	0.76	0.58
BIG	Largest diameter tree	20	0.13	0.02	0.35	0.12
VIT	Average diameter of *Vitex pinnata*	13	0.70	0.49	0.76	0.58
COP	Average diameter of all coppiced trees	12	-0.26	0.07	0.20	0.04
FRD	Average diameter of all fruit trees	15	0.44	0.19	0.61	0.37
Kian	Average diameter of *Artocarpus odoratissimus*	3	0.31	0.10	0.76	0.58
Bungau	Average diameter of *Artocarpus lanceifolius*	4	0.15	0.02	0.94	0.89
Jilen	Average diameter of *Dimocarpus longan*	3	-0.73	0.53	0.67	0.44
Lasiu	Average diameter of *Lancium domesticum*	2	1.0	1.0	0.99	0.97
Meu	Average diameter of *Nephelium ramboutan-ake*	10	0.66	0.44	0.79	0.62
Pangin Alo	Average diameter of *Mangifera indica*	3	0.92	0.84	0.98	0.95
FUR	Height of first fork over total tree height	20	0.15	0.02	0.31	0.09
HGT	Average height of all trees	20	0.26	0.07	0.47	0.22

[a]Includes a fictional plot of zero age ($*N = N + 1$), for which values for all variables are equal to zero.

whereas the largest tree and average height of the canopy were very poor predictors.

Analysis of the Vegetation of Abandoned Camps

This section presents the results of several statistical tests chosen to compare the vegetation of the twenty sample plots. The aim is to see if there are lasting changes in floristic structure and diversity resulting from Penan settlement and activities at the campsites.

STAND STRUCTURE. The first set of tests compared only the stand structure variables (see table 1.2), which are all continuous and therefore could be analyzed using cluster analysis (fig. 1.4) and principle component analysis (PCA; see fig. 1.5 and table 1.7).

At first glance, the clusters of sample plots in fig. 1.4 appear to reflect differences in age since abandonment of the sites (ABN) and the current forest type as classified by the Penan (VEG). The more frequently used and thus more disturbed vegetation types (such as the Kenyah Badeng orchards) are clustered at the top of the diagram (D, C, and S). The older and seemingly less disturbed plots in mature and primary forest are in two clusters at the bottom (H, M, K, J, L and

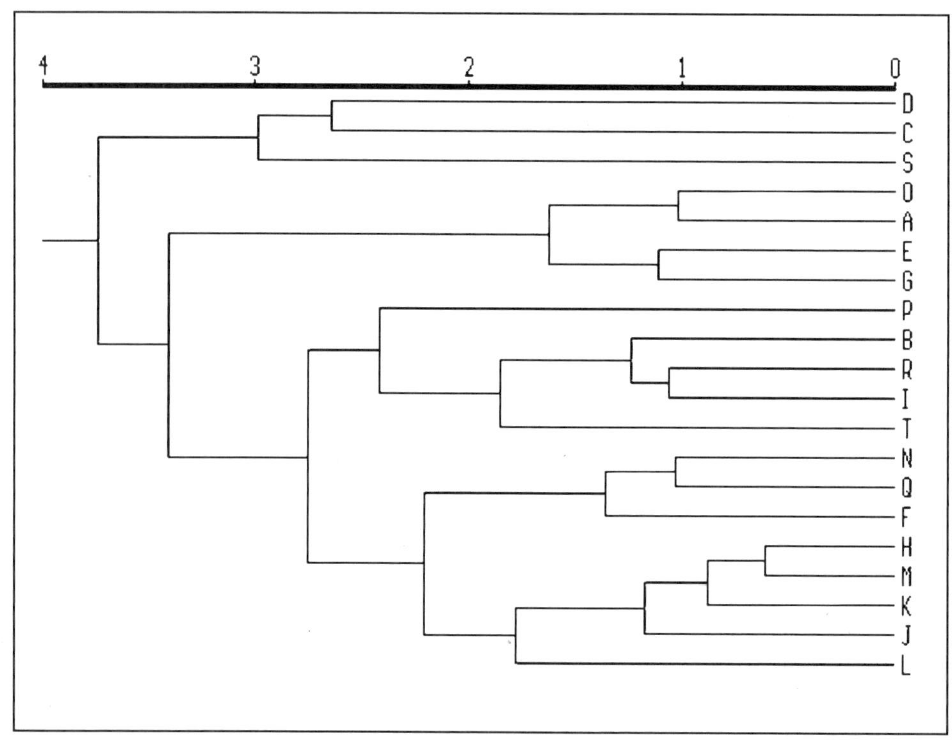

Figure 1.4 Phenogram of Peliran sample plots from cluster analysis of stand structure variables ($r = 0.71$). See table 1.2 for an explanation of variables and for a description of sample plots.

N, Q, F). Between these two extremes are clusters of what appear to be sites in old (O, A, E, G) and young secondary forests (P, B, R, I, T). One might therefore assume that variation in stand structure simply reflects forest type and that the effects of manipulating vegetation in site construction and settlement are lost over time, such that older sites come to be unrecognizable from undisturbed primary forests. If that were the case, there would be no long-term effect of Penan campsites on the structure of forest vegetation.

However, when examined closely the clusters have a varied mix of sample plots from different forest types. One would expect a Kenyah fruit orchard, O, to be in the first cluster and a former Penan fruit camp in old secondary forest, C, to be in the second cluster. Also, camps N, Q, and F, are all from different parts of the territory of Long Peliran, and from different forest types, yet they are clustered together. Likewise, the cluster of camps O, A, E, and G are similar in

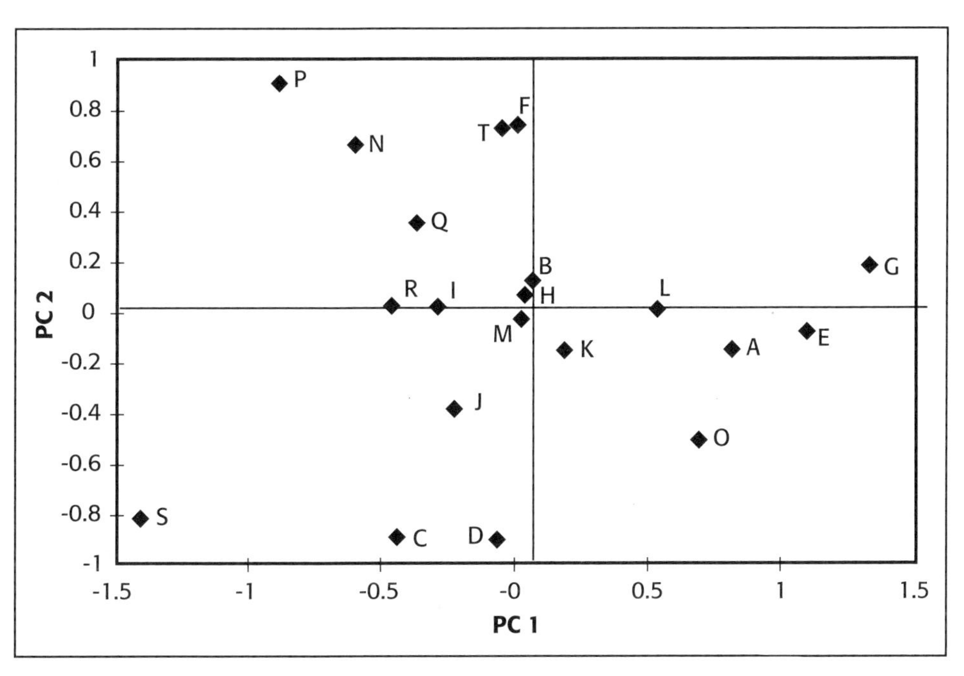

Figure 1.5. Projection of Peliran sample plots in the space of the first two components according to analysis of stand structure variables.

Table 1.7 *Variation explained by the first three principle components and the stand structure variables with greater weight in each component*

Principle Component	Variation Explained (%)	Accumulated (%)	Variables with Higher Loadings
PC 1	43.35	43.35	DEN (+), BAS (+), HGT (+)
PC 2	27.60	70.95	DBH (-)
PC 3	24.5	95.46	FUR (+)

Note: Signs in parentheses indicate positive or negative relationships among variables. In figure 5, PC 1 variables increase from left to right on the *x*-axis, while the PC 2 variable decreases from top to bottom along the *y*-axis. PC 3 is not displayed in figure 5.

their age and the presence of tall trees, but differ in their floristic composition and their history of associated human activities and therefore are classified differently. These results suggest that the physical attributes of a forest, described by stand structure variables, are not good predictors of the forest type, as defined by the Penan. It is possible that the Penan have made mistakes in classifying the forest in the sample plots or that these variables are not the sole criteria used in the Penan classification system. More likely, human manipulation of the vegetation at these sites is responsible for altering the structure of the forest and thus blurring the boundaries between forest types.

These alternative explanations can be addressed, in part, by the principle component analysis of stand structure variables. The scatter diagram of sample plots in fig. 1.5 shows groups similar to those derived from the cluster analysis, while table 1.7 identifies those structural characteristics that serve to explain the variation in clustering of sample plots.

The first component (or dimension) shows correlation between density of trees (DEN), total basal area of trees (BAS), and height of trees (HGT), increasing from left to right in fig. 1.5. The average diameter of trees (DBH) increases vertically from top to bottom in fig. 1.5. Thus, plots in Kenyah orchards are clustered because they are relatively open forest areas with a few large but short trees. This appears to be typical of older orchards managed regularly for the production of fruit. As in the cluster diagram, sample plot C is included in this group, even though it is classified as old secondary forest, because it has begun to resemble an orchard. Local history and an examination of floristic diversity, described later, support the notion that this camp was established in an old Kenyah swidden garden—often the precursor of a fruit grove or orchard. Since then, the site has been visited frequently by the Penan and Kenyah and apparently has been subject to management activities for the production of fruit. Orchards that have been neglected or abandoned begin to resemble site O, which shows a higher density of trees (including more nonfruit trees) that are narrower and taller, suggesting that it has not been subject to recent management activities such as pruning, selective felling, or burning. Young orchards, such as site C, have a lower furcation index (FUR), indicating that pruning has occurred more recently than in sites S, D, and O.

Camp sites established in swiddens or at the edges of swiddens (e.g., P, N, T, R, I) are similar in structure to the orchards in having a few short trees, but these trees are smaller in diameter and bifurcate much closer to the ground, suggesting a younger age. These sites are also subject to management activities that may retard normal regrowth. In contrast, site F—an old sago-processing camp next to a small stream and just below a ridge-top grove of hill sago *(Eugeissona utilis)*—may demonstrate regrowth in a less manipulated site. Structurally, the older sago camp appears to resemble an old abandoned Kenyah swidden site, except that it is denser and has taller trees. While it is possible that the Penan may have misunderstood the history of the site, the composition of the regrowth appears to be consistent with that found in primary forest (see fig. 1.6). As a daily work camp, the site was probably cleared of smaller trees in order to build a resting platform and an area for pounding the palm pith but was not subjected to the kind of intensive management activities found in more permanent sites. On the other hand, the forest growing on the site today is noticeably different from the surrounding primary forest, evidence that even a limited degree of management nearly thirty years ago can have a lasting impact.

Finally, campsites in mature and primary forest (H, J, L, K, and M) have average densities and tree sizes, medium to tall trees, and high furcation indexes. All of these sites are located on ridge tops, and many are alongside seldom-traveled trails. In contrast, camp G, a two-hour walk from Long Peliran, lies across a major thoroughfare used by hunters to access primary forest and sago palm groves—those above camp F mentioned earlier. This site was occupied around thirty years ago and unlike other camp sites in mature forest it has a surprising combination of high tree density, thin but tall trees alongside the trail, and a low furcation index (see fig. 1.5). The high density and height of its trees remain mysterious but may be due to the compatibility of the local and introduced flora with microhabitat variations along the ridge in soil nutrients and moisture and protection from high winds. Included in the introduced flora are fruit trees, including an abundant species of jackfruit, *bungau* (*Artocarpus lanceifolius* Roxb.), that date back to the Penan settlement and continue to be an attraction for passing travelers. The fruit from these trees is harvested annually by means of pruning and felling. The low furcation index for the site may result from this continuous management by visitors. Thus, long after its initial abandonment, site G retains a distinct managed vegetation that is in part a result of Penan settlement and subsistence activities.

DIVERSITY AND DISTURBANCE. The diversity variables examined were described in the methods section (table 1.3). Results of regression analysis between these diversity variables and several ecological and anthropological variables of the sample plots are presented in table 1.8. As expected, tree species richness (MGT) roughly decreases as human manipulation of the vegetation (VEG) increases (R=0.82); thus, the older sites in primary forest had a higher number of tree species than those of the older managed orchards and younger swidden camps. This relationship has its spatial correlate as well, whereby managed sites tend to be closer to the Kenyah village while primary forest sites tend to be farther away. It is not surprising, therefore, that the analysis shows that tree species richness increases in sites at higher altitudes (ALT) and at greater distances (DIS) from Long Peliran. The same relationships hold when we examine taxonomic abundance (FST). As with the data for trees, the diversity of saplings in the sample plots (MGS and FSS) increased with older, less managed forests (VEG) but was not strongly influ-

Table 1.8 *Results of linear regression analysis (Pearson's R values) between diversity variables in rows and ecological and anthropological variables in columns*

	ALT	AGE	DIS	VEG
MGT	0.86	-0.08	0.44	0.82
MGS	0.56	-0.05	0.48	0.61
FST	0.60	-0.09	0.50	0.52
FSS	0.55	-0.09	0.40	0.65
FRT	-0.62	0.68	-0.64	-0.77
FDO	-0.59	0.71	-0.56	-0.71
VDO	-0.42	-0.50	0.13	-0.43

Note: See tables 1.1–1.5 for explanation of variables.

enced by altitude, distance from the village, or the other disturbance variables.

Not surprisingly, dominance of a single species (DOM) decreases as species richness increases. *Vitex pinnata* was dominant in recent campsites in or adjacent to former rice swiddens. The dominant species was a fruit tree in the managed orchards, fruit camps, and a few older sites, while primary forest trees (e.g., oaks, laurels, and dipterocarps) dominated camps in mature or undisturbed forests. Fruit diversity (FRT) and dominance (FDO) appear most highly correlated with forest type (VEG), again increasing in managed forests (table 1.8). However, camp B and C have a higher fruit dominance than expected given the surrounding secondary forest. In fact, these "islands of fruit" are located in old swidden gardens that once surrounded temporary shelters and gardens. Repeated management during Penan and Kenyah fruit-collecting expeditions has served to maintain the fruit trees over the years. Site O, also discussed earlier, has fewer fruit trees than the more actively managed orchards. Interestingly, *Vitex pinnata*—considered a good indicator of previous swidden fields (as in sites P, N, R and T)—is more abundant in this former orchard than in the two old swidden garden sites, B and C.

Because *Vitex* is a high-quality fuelwood, it may be repeatedly pruned or felled by visiting fruit collectors or nearby residents. Thus, the more frequently used fruit-collecting sites should show a reduced number of these trees, a higher percentage of coppiced trees, and a lower furcation index. This is exactly the case when comparing camps B and C to site O. Thus, it seems likely that all three of these sites were swidden fields before they were planted with fruit trees and then subjected to varying degrees of human management.

While the sample sizes are small and most of the tests not statistically significant, these results at least suggest that the species richness of trees and saplings in Penan campsites and Kenyah orchards has been influenced by the intensity and duration of disturbances caused by human activities at these sites. The activities identified so far include campsite preparation, collection of fuelwood, swidden cultivation and burning, and pruning and felling of fruit trees.

FLORISTIC COMPOSITION. The third set of statistical tests went beyond measures of species richness and compared the abundance and distribution of tree species across the twenty sample plots. These tests included both cluster and correspondence analyses of floristic composition, measured by proportional abundance of tree taxa (dbh >10 cm; see figs. 1.6 and 1.7, and table 1.9).

The phenogram in figure 1.6 is noticeably different from that in figure 1.5, which displays the results of the cluster analysis of stand structure variables. The first division (moving down the diagram from left to right) differentiates those sites located in primary forest from the rest. The next divisions are more complicated, with several loose clusters of a few sites distantly related to seemingly unique sites. There is a cluster of two camps in secondary forest, I and A, with site N distantly related. All three camps have few or no fruit trees. Kenyah orchards and Penan camps in swidden fruit groves (B, D, O, and S) all cluster together, except for site C. Camps in younger swidden fallow (P, R, and T) form a cluster with Site E—in very old swidden fallow—as an outlier of that group. Sites in mature forest (G and H) form a loose cluster. Site Q is an outlier to these last two clusters. Generally speaking, the phenogram suggests that few sample plots are very similar. It highlights the diversity of assemblages of

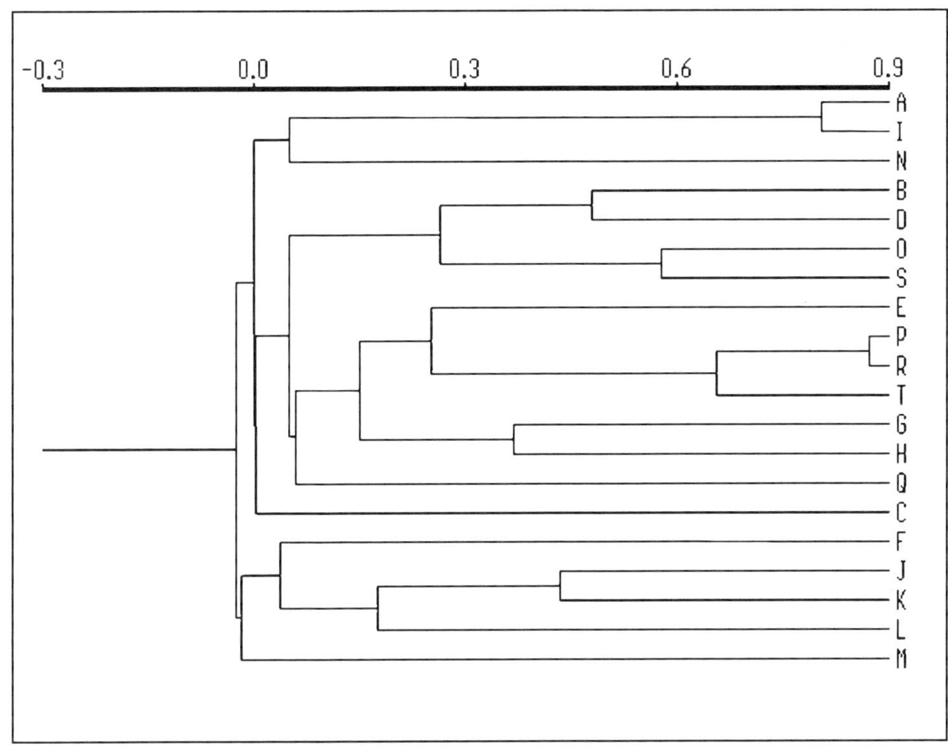

Figure 1.6 Phenogram of Peliran sample plots from cluster analysis of floristic composition. The cluster analysis correlated the sample plots using the proportional abundance of tree taxa (dbh > 10 cm). The fit of the phenogram to the correlation matrix is high ($r = 0.866$).

tree flora in the old campsites but tells us little about the kinds of factors or dimensions underlying these groups and outliers.

Correspondence analysis of this same data set reveals that fruit tree abundance drops as primary tree species increase in abundance in campsites (from upper left to lower right in figure 1.7). Table 1.9 shows the tree taxa that differentiate the sites. The first factor, separating sites along the x-axis in figure 1.7, shows an increase in primary forest tree species as one goes from left to right. Tree species with a negative first factor in table 1.9 are secondary forest species that increase in relative abundance from right to left. The second factor, separating sites along the y-axis, shows an increase in

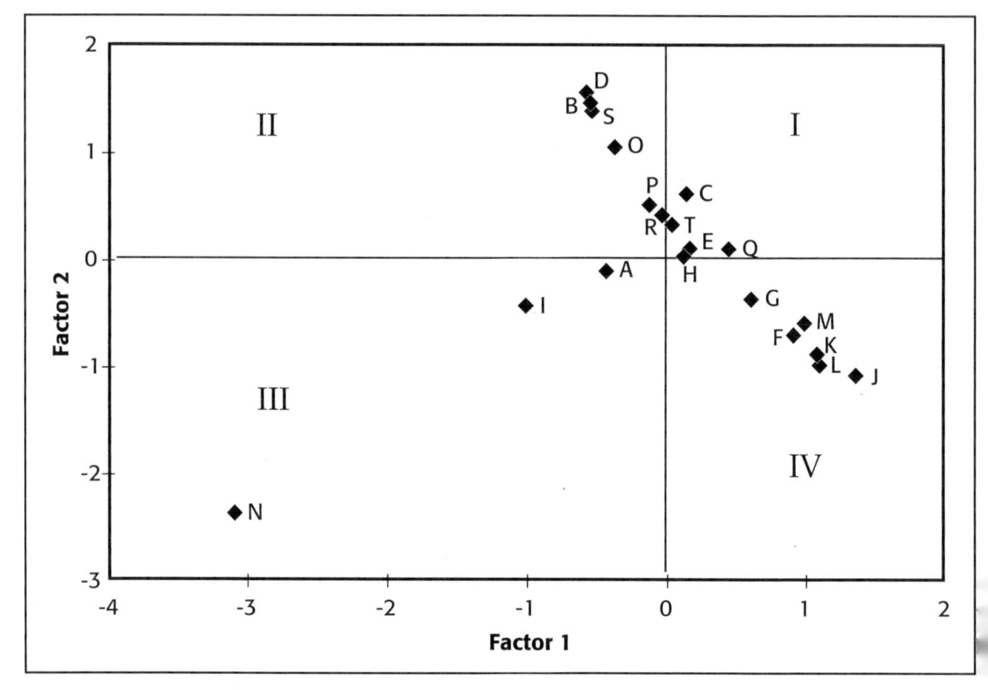

Figure 1.7 Projection of Peliran sample plots in the space of the first two factors, or dimensions, from correspondence analysis of floristic composition. The analysis compared the proportional abundance of tree taxa (dbh > 10 cm) in the sample plots. The clusters closely match those described in the cluster analysis in figure 1.6.

the abundance of fruit tree species from bottom to top. As was suggested by the cluster analysis, outlier sites N, I, and A show lower than expected abundance of fruit given their abundance of secondary forest species, which might be due to the consequences of swidden cultivation. Thus, looking at the distribution of sites in figure 1.7, one gets the impression that the impact of Penan fruit collecting and fruit camps on floristic composition diminishes over time, as the oldest sites are those in mature and primary forests, which have very few fruit trees. Does this mean that floristic composition, at least for trees, is a better predictor of forest type, and by extension human influences, than the stand structure or diversity variables discussed above? Which variables, acting alone or together,

Table 1.9 *Factor loadings for Penan tree taxa from correspondence analysis, as displayed in figure 1.7*

Factor 1	Factor 2	Penan Taxa	Species	Forest/Use	Plots
1.406618	-1.13525	tepusang	*Garcinia maingayi*	Fruit	J
1.406618	-1.13525	temeu bura'	*Memecylon costatum*	1st forest	J
1.406618	-1.13525	nye asu	*Meliosma sumatrana*	1st forest	J
1.406618	-1.13525	meu tai ayau	*Nephelium* sp.	Fruit	J
1.406618	-1.13525	leumek	*Mastixia trichotoma*	1st forest	J
1.406618	-1.13525	leran	*Dillenia* sp.	1st forest	J
1.406618	-1.13525	apu lodat	*Baccaurea stipulata*	1st forest	J
-2.66672	-1.94622	tegolem talun asi	*Vitex pinnata*	2nd forest	I,N
-3.11549	-2.42602	menuang	*Macaranga caladifolia*	2nd forest	L,N
-3.22608	-2.46221	polu	*Aglaia* sp.	2nd forest	N
-3.22608	-2.46221	metekang	*Licania splendens*	2nd forest	N
-0.56417	**1.504661**	jilen	*Dimocarpus longan*	Fruit	B,D,O
-0.56874	**1.540443**	basut	*Artocarpus odoratis-simus*	Fruit	B,D,M
-0.57478	**1.52655**	paken	*Durio zibethinus*	Fruit	B
-0.57478	**1.52655**	lieng	*Aporusa nitida*	Fruit	B
-0.57478	**1.52655**	keramo	*Dacryodes rostrata*	Fruit	B
-0.60778	**1.622553**	lasiu	*Lansium domesticum*	Fruit	B,D
-0.60978	**1.628365**	pangin	*Mangifera pajang*	Fruit	D

are responsible for the similarities and differences of the sample plots? Untangling the importance of the anthropological and biophysical variables at work here requires one final analytic step, a principle coordinate analysis of all the qualitative and quantitative variables analyzed so far (fig. 1.8, table 1.10).

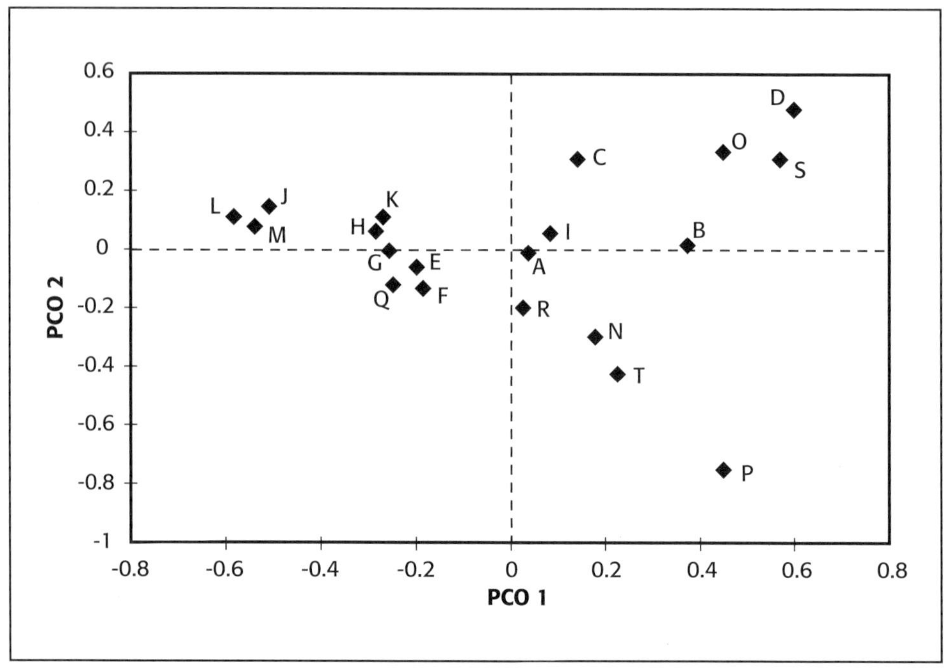

Figure 1.8 Plot of Penan forest camps resulting from a principle coordinate analysis according to twenty-four ecological and anthropological variables

Results of the principle coordinate analysis, comparing the sites using both quantitative and qualitative variables, are presented in table 1.8 and figure 1.8. The first component shows high positive correlation between the age of sites, altitude, and tree species diversity, while tree size and fruit dominance are positively correlated with each other and negatively correlated with the first three variables. This means that among the sample plots older sites were located at higher altitudes, had a greater species diversity, had smaller average tree diameters, and had less dominance by fruit trees (in quadrant II). This is what one would expect at older sites that have been left to regenerate naturally; the effects of the initial manipulations are not lasting. The second component shows a high positive correlation among variables measuring human disturbance. This component differentiates camps that have had varying degrees of human disturbance, primarily the difference between fruit gardens

Table 1.10 *Factor loadings from principle coordinate analysis for quantitative and qualitative variables describing sample plots in Long Peliran, as displayed in figure 1.7*

Variable	PCO 1	PCO 2	PCO 3
VEG	0.494[a]	0.039	0.020
FRT	-0.617[a]	-0.379	-0.026
FDO	-0.598[a]	-0.452	-0.051
VIT	-0.337	-0.057	0.064
VDO	-0.433	[a]0.491	-0.072
COP	-0.236	[a]0.492	0.262
CDO	-0.406	[a]0.514	-0.007
DEN	0.177	0.058	[a]0.513
BAS	-0.114	-0.210	[a]0.589
DBH	-0.499[a]	-0.477	-0.032
HGT	0.146	-0.400	0.281
FUR	0.217	-0.314	-0.368
MGT	0.470[a]	-0.136	0.052
MGS	0.368	-0.123	-0.228
FST	0.237	-0.087	-0.163
FSS	0.350	-0.134	-0.268
ALT	0.458[a]	-0.026	0.016
DIS	0.197	0.171	-0.319
TOP	0.394	0.082	0.191
ACT	-0.237	0.154	0.040
OCC	-0.142	0.057	-0.291
TIM	-0.242	0.207	-0.293
ABN	0.322	0.171	0.200
POP	0.031	0.358	-0.110

[a]Highly correlated variables.

(in quadrant I) and swiddens (in quadrant IV) but also between older sites that have been subject to longer lasting or more recent disturbances (in quadrant III). The third component seems to be related to stand structure, particularly tree density, which varies among sites that contain or are near walking trails. These are conditions found in the managed village orchards and a couple of the Penan camps that were established in old swidden fruit groves.

Again, these results support the view that tree species diversity of campsites is very much a consequence of the intensity of human manipulations of the forest environment. For younger sites, the effects of the Penan camps in the mature tree vegetation are invisible, and older camps—in tall, mature forest on the periphery of the farmers' influence—show little lasting evidence of Penan settlement over time. Instead, the results suggest that at the recent sites tree diversity may be more a consequence of Kenyah farmers' forest management practices prior to the establishment of campsites by the Penan.

SAPLINGS. On the other hand, the diversity of sapling and seedling vegetation is very much the unintentional result of Penan activities. Many of those sites that had few fruit trees had a greater abundance of fruit saplings (fig. 1.9).

The diversity and abundance of fruit species in the sapling vegetation was high in forest fruit camps and village orchards and decreased in older and less manipulated sites. Sapling species richness was not strongly correlated with tree species richness ($R^2 = 0.69$) nor with any other variables tested in this analysis, suggesting that sapling diversity is different at every site and may in fact be dependent on complex interactions of site-specific historical-ecological conditions.

Summary of Results

The results may be summarized as follows:

1. The Penan who migrated into the Long Peliran area in the 1950s experienced a period of transition beginning around 1965, whereby they moved closer to and eventually into the Kenyah village. The increase in proximity to the farmers corresponds to a decrease in the altitude of Penan settlement sites (see table 1.11). The rate of movement seems high; over a

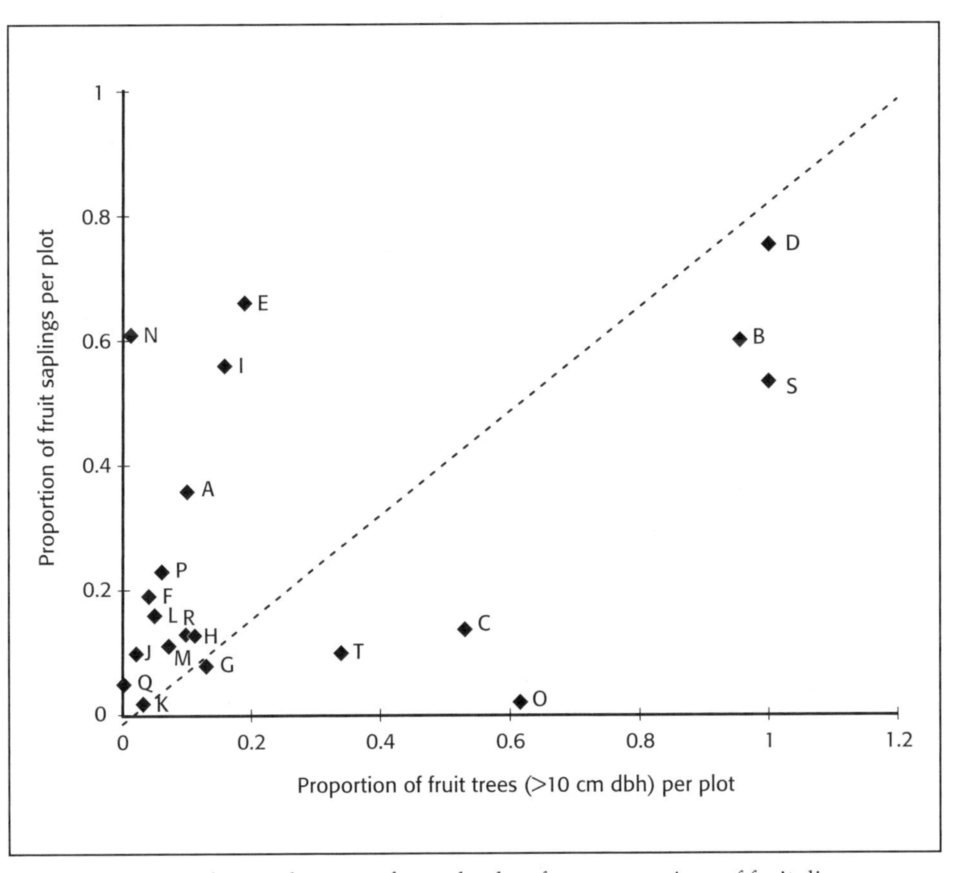

Figure 1.9 Scatter diagram of sample plots from comparison of fruit diversity in canopy (x-axis) and sapling layer (y-axis). Dotted line indicates 1:1 ratio.

thirty-year period, one of these families moved fifty-one times among thirty-one different campsites.

2. The change in settlement pattern was accompanied by a shift in subsistence strategy. As a whole, the group began to adopt farming practices similar to those of their neighbors, including rice swidden cultivation and gardening of crops such as manioc, bananas, taro, and corn. They are continuing the traditional practices of hunting, processing sago palms,

Table 1.11 *Changes in Penan campsites since the 1960s*

	Older Campsites (pre-1965)	Newer Campsites (post-1965)
Original forest type	Primary forest, natural secondary forests	Old swidden fallow, cultivated lands, farmer's villages
Altitude	High	Low
Distance to Village	Far	Residence sites close, expedition camps far
Topography	Ridgetops, small streams	Near rivers and large streams
Size	Large forest camp with small forest satellite camps	Large village site with small forest satellites
Site activity functions	General	Specific (farming added)
Frequency of reuse	Infrequent	Frequent

harvesting fruit, and collecting forest products for trade, but the mix of these practices is now more varied among the group's families and often differs from year to year.

3. Whereas settlement sites once could be found almost anywhere in the forested mountains, the recent period has seen the establishment of Penan villages on the banks of the larger rivers and settlements at new swidden fields or in Kenyah villages. Only small satellite camps, established for hunters and forest product collectors, are now found in the headwaters area. Usually within an hour or two of the new settlements, Penan fruit camps are occupied for weeks at a time during the annual fruit season. Most of these are located in abandoned Kenyah swidden gardens. Kenyah farmers planted the large fruit trees now harvested by the Penan. Few older sites, in primary forest, appear to have been occupied specifically for fruit collecting.
4. Manipulation of the vegetation at a campsite occurs as a result of clearing a site, constructing a shelter, provisioning the

residents with food and fuel, and conducting work-related activities. Trees may be felled, pruned, debarked, or scarred. Sapling trees are often felled. Undergrowth may be cleared or burned. Vines, fruit, edible palm leaf buds, ferns, mushrooms, bark, large herbaceous leaves, and standing and fallen deadwood may be harvested or collected nearby.

5. Old campsites are often found next to current ones and may be identified by groves of fruit trees, a clearing with coppiced and scarred trees, or piles of rocks. The Penan identify *Mangifera pajang*, *Lansium domesticum*, *Dacryodes rostrata*, *Durio zibethinus*, *Artocarpus odoratissimus*, *Nephelium ramboutan-ake*, and *Dimocarpus longan* as excellent indicators of past human presence, especially if they are found in close proximity to each other. Research confirmed this correlation, as all surveyed campsites contained some of these species in the tree or sapling flora.
6. Descendants' oral histories and reconstructed genealogies were used to estimate the dates of occupancy and age of forest vegetation in and around the campsites. Dating sites using tree growth rates proved to be difficult because of the incongruity of forest and human settlement history. The average diameter of all trees (dbh > 10 cm) in a sample plot proved to be the best predictor of the age of a campsite's vegetation, whereas the largest tree diameter and average height of the canopy were very poor predictors. The average diameter size of fruit trees, *Vitex pinnata*, and all coppiced trees was positively correlated with the estimated age of the vegetation, albeit at statistically low levels of significance.

The analysis of the vegetation in and around 18 Penan camps, and for comparative purposes two Kenyah orchards, produced the following results.

1. The Penan campsites varied significantly in terms of the surrounding vegetation. One was located in a village orchard (S), one was found in young swidden fallow (P), two were located in abandoned Kenyah swidden fruit groves (B, C), several were located in older Kenyah swidden regrowth (A, E, I, N, R, T),

and the remainder were found in mature forest (F, G, H, J, K, L, M, Q).

2. Kenyah orchards were the most manipulated forest type examined (sites D, O, and S). Owners regularly harvest fruit by either pruning branches or felling trees, the undergrowth is often trampled or burned, and non-fruit-producing trees such as *Vitex pinnata* are eventually cut out. Orchards that are tended typically have a lower canopy height, fewer trees, and a lower furcation index. This study found that the average diameter of trees in orchards is greater than those in other forest types either as a result of their older age or, more likely, due to the younger, smaller trees being cut out. In general, campsites established in swiddens or at the edges of swiddens are similar in structure to the orchards in having a few short trees, but these trees are smaller in diameter and bifurcate much closer to the ground, suggesting a younger age. Campsites in mature and primary forest have average densities and tree sizes, medium to tall trees, and high furcation indices.
3. Vegetation structure varied among the campsites of each forest type. Thus, the physical attributes of a forest, described by stand structure variables, are not perfect predictors of the forest type, as defined by the Penan. It is possible that the Penan have made mistakes in classifying the forest in the sample plots, or that these variables are not the sole criteria used in the Penan classification system. More likely, human manipulation of the vegetation at these sites is responsible for altering the structure of the forest and thus blurring the distinction between forest types. Thus, camps B and C resemble the orchards in terms of vegetation structure, while the recently neglected orchard O has begun to resemble the very old swidden forest of camps A and E. The vegetation in campsite G, in primary forest, is subject to frequent disturbance due to its proximity to a trail and also resembles old swidden forest. The forest in the sago-processing site, F, appears to be old but natural secondary forest, structurally more similar to old swidden forest (e.g., N and T).

4. Tree species richness roughly decreases as human manipulation of the vegetation increases. Thus, orchards had the highest proportion of fruit trees and fruit saplings—evidence of management—but very low species richness and taxonomic abundance for both trees and saplings. *Vitex pinnata* was found to be dominant at recent campsites in or adjacent to former rice swiddens, which also showed low species richness. As orchards and other managed sites tend to be closer to the Kenyah village and primary forest sites tend to be farther away, it follows that tree species richness increases in sites at higher altitudes and greater distances from Long Peliran as a result of human manipulation of the vegetation.
5. Analysis of floristic diversity highlights the diversity of assemblages of tree flora in the old campsites. Fruit tree abundance drops as primary tree species increase in abundance at campsites. Thus, the impact of Penan fruit collecting and fruit camps on floristic composition appears to diminish over time, as the oldest sites are those in mature and primary forest, which have very few fruit trees. Floristic composition appears to be a good predictor of forest type and by extension the effects of human manipulation of the vegetation. The principle coordinate analysis suggests that older sites were located at higher altitudes, and had greater species diversity, a smaller average tree diameter, and less dominance by fruit trees. This is what one would expect in older sites that had been left to regenerate naturally; the effects of the initial manipulations are not lasting.
6. The diversity and abundance of saplings of fruit species was high in forest fruit camps and village orchards and decreased in older and less manipulated sites. Seemingly contradictory to this, the proportion of fruit tree saplings to fruit trees was much higher in the older sites and much lower in the manipulated forest types. This suggests that fruit tree species could very well dominate the next generation of trees in the older sites.
7. Anthropological factors, such as group size, the number and length of occupations, and the primary function of the camp,

did not prove to be statistically significant in determining the present floristic diversity of the camps, although some were important for understanding the vegetation in particular campsites.

Discussion and Conclusion

The findings of this study are preliminary and, though inconclusive in many respects, may be used to address a number of substantive, theoretical, and methodological issues raised at the outset of this chapter regarding human-environment interactions in tropical forests.

Consideration of the history of human-forest interactions proved to be critical in arriving at the following conclusion: the effects of Penan settlement on the tree vegetation of campsites are practically invisible in younger sites and only minor in older ones. Instead, the structure and diversity of tree flora is more a consequence of Kenyah farmers' forest management practices prior to the establishment of campsites by the Penan. Through reoccupation of the Kenyah sites, the Penan perpetuate the effects of the farmers on the forest vegetation. One consequence of the continued use of a site is the maintenance of a permanent clearing and a corresponding reduction in the number of saplings, especially fruit saplings. Thus, managed orchards and fruit groves have relatively fewer fruit saplings than those at more remote, less frequently visited sites. In the latter cases, the more or less abandoned sites, the proportion of fruit saplings is greater than that of fruit trees. If those surviving fruit saplings eventually become established in the tree canopy, forming a fruit grove, then this will be evidence of a significant long-term effect of Penan activities. Unfortunately, a campsite of this type has yet to be studied, or even located, anywhere in the Lurah river valley. The ubiquity of former farming villages, old swidden regrowth, and fruit groves makes it unlikely that any such sites will be found here. Since the turn of the century, the Penan have been living in an old anthropogenic landscape.

Central Borneo in general has a long history of human settlement with frequent long-distance migrations, intertribal warfare, extensive agriculture, and widespread collection of forest products for trade (see Brookfield, Potter, and Byron 1996). There is little ground that has never been tread upon, and most human settlement areas have been occupied or visited repeatedly; pristine forest and truly abandoned sites are very rare indeed. Given such complications in site history, it is not surprising that a major hurdle that arose during the course of this study was how to differentiate regrowth resulting from Penan occupations as opposed to any previous human activity. Failure to do so would undoubtedly lead to misinterpretations of the age of abandonment and the age of the present day vegetation, making it all but impossible to distinguish the consequences of the actions of different human groups. Fortunately, the camps surveyed were all young enough to have former occupants still living. These Penan and Kenyah elders were able to locate the sites, describe past site functions, and eventually piece together a history of land use and settlement for the village and surrounding territory during the last seventy years. Still, there are unexplored factors that may have had a significant impact on the environmental history of the village. Most importantly, the Pujungan area was occupied by Kayan people prior to the arrival of the Kenyah in the 1850s and the Penan in the 1890s. In fact, there was a Kayan longhouse at the mouth of the Lurah, just across the Bahau River from the present village of Long Peliran. Kenyah elders point to large fruit trees near the riverbank and claim that they are the last of the Kayan fruit orchards, but other than those trees the consequences of Kayan settlement for the area's vegetation are largely unknown.

Whether it is the Bornean tropical forest—rich in species—or an isolated oceanic island like Hawai'i—high in endemics—the ecological conditions observed today are the consequence of the complex historical interaction of local and extralocal factors. These factors include climate, geology, biogeography, other biophysical factors, and human ecology—in the broadest sense of the term—including social, cultural, economic, and political aspects of human interaction with the environment at local, regional, and global levels of organization (see Caldecott 1992; Brookfield, Potter, and Byron 1996). For instance, in a study of the prehistory of the Amazon area, Roosevelt et al. (1996, 382) found that

> forests once thought to be virgin were settled, cut, burned and cultivated repeatedly during prehistoric and historic times, and that human activities widely altered topography, soil and water quality. ... Substantial biodiversity patterning appears to be associated with such human activities. ... Given the nature of the human occupation, it seems reasonable to acknowledge a human role in the development of landforms and biotic communities in Amazonia over the millennia. Research on tropical forests can benefit by taking into account the long-term effects of both past and present human activities.

Indeed, Balée (1989) estimates that 12 percent of the Amazon is anthropogenic. A number of studies have already been conducted along the lines recommended by Roosevelt et al. in various subfields of anthropology and conservation biology.[7] This research has shown how tropical forest farmers, past and present, manipulate vegetation communities through simple management techniques such as pruning; weeding; protecting desired individuals; intercropping in fallow swiddens; transplanting wild species into gardens, orchards, and degraded areas; and preserving varieties of staple cultivars. The aims of these manipulations are to increase the abundance and concentration of important biological resources in and around human settlements, in both the spatial and temporal dimensions and thereby reduce the costs and uncertainty of the production of biological resources. The consequences of such human actions for the structure and function of tropical forests are profound, amounting to an ordering of a chaotic landscape: a cosmos of human-influenced floristic patches, soils, and faunal distributions. Moran (1993, 761), discussing upland forests in the Amazon, states that "on the better soils one finds anthropogenic forests which concentrate species of economic value without excessive simplification of species in the ecosystem, so much so that for a long time we thought of them as 'virgin' forests." Other studies indicate that some groups, including the Kenyah and Penan of the Kayan Mentarang area, can coexist with high levels of biodiversity and in some cases may even be responsible for those levels.[8]

While it is known that the adoption of agriculture and the domestication of animals led to greater conscious manipulation of soils, geomorphology, flora, and fauna for human ends, the long-term consequences of the activities of nonagricultural societies, where the impact on flora and fauna is believed to have been minor or largely unintentional, are still poorly understood (see Brosius 1991; and Dwyer and Minnegal 1991). Outside of academic debates, hunter-gatherers are portrayed typically as living harmoniously with nature (Balée 1998; Bodley 1990; Headland 1997; Sponsel 2001, Redford 1990). Such veneration is evident in contemporary New Age spiritualism and Nature worship (Ellen 1986; Soulé 1995). The trend to use the traditional knowledge and management systems of hunter-gatherers and other indigenes as resources, even potential models, in conservation and development projects may also be derived from essentialist views of natives as conservationists (Agrawal 1995; Eghenter 2000). Unfortunately, there is a tendency among biological ecologists to either ignore humans, as if they were unnatural contaminants of pristine nature, or focus on them as "agents of environmental degradation, resource depletion, habitat destruction, and species extinction" (Sponsel 1997a, 143). Not surprisingly, their research tends to demonstrate that humans, and in particular indigenous peoples, are not the noble conservationists that the West wishes them to be (Redford 1990; Heinen and Low 1992; Alvard 1993). Reconstruction of human-environment interactions over long time spans, primarily from archaeological and historical evidence, suggests many cases in which humans have had negative, even disastrous, effects on their environments (Martin 1984; Clay 1988; Bodley 1990; Low and Heinen 1993). They have caused the extinction of rare and endemic species (e.g., endemic birds in Hawaii, see Olson and James 1984), important food sources (e.g., the big game hunters of Pleistocene North America; see Martin 1984), and habitats (e.g., Maya rain forests in Mexico; see Turner 1982). On the other hand, Moran (1993, 754) writes:

> The view that native peoples of Amazonia are "backward" (a view common in national government circles) must be replaced by appreciation of their stewardship of that region. The view that they are "noble savages" is no less off the mark. They vary a great a deal amongst themselves. Nevertheless, the imprint they have left on the forests has been far less destructive than other, non-Amazonian natives despite their use of the region for millennia, compared to the impact that others have had in comparatively small number of years.

As illustrated in this study of the Penan, there are currently few ways to track hunter-gatherers through time, partly because evidence of their settlements, daily activities, and even burial sites is difficult to find in the tropical forest. They also leave few nonperishable remains behind and never stay in one place for long. There are the famous prehistoric caves sites, such as those at Niah or, closer to the Kayan Mentarang, Madai in Sabah (Bellwood 1989), which tell us something of the subsistence patterns of early Bornean hunter-gatherers, including what kinds of animals were captured and eaten. But evidence of the use of plants is rare or can only be implied, and the effects of hunting and gathering on plant and animal populations at that time are largely unknown. More recently, in the twentieth century the Penan along with many others hunted the Sumatran rhinoceros to near extirpation on Borneo.

When fruit trees are planted, or preexisting groves perpetuated, the Penan can have dramatic and visible effects on the forest ecosystem. Although comparative data are lacking at this time, ongoing research at a variety of undisturbed and disturbed study sites indicates that the Penan affect the abundance and distribution not only of flora but fauna. Fruit groves appear to be important islands of food resources for animals, especially during off-mast fruiting seasons when dipterocarps and other trees do not fruit. Thus, Penan regularly hunt in these campsites, confident of finding animal prey feeding on the fruit. The effects of the much larger abandoned Kenyah villages on forestwide fruit production and animal foraging habits are probably even more significant (Puri 1996). Other than the planting of fruit trees, two of the more consequential subsistence practices, still carried out today, that probably have had significant and visible effects on the Bornean forest ecosystems are clearing with fire and the collecting of sago palms. Similar to groups of Australian Aborigines and North American Indians, the Penan have maintained grassland habitats for preferred prey through seasonal burning of dense stands of ferns *(Dicronopteris linearis)* or dry "cogon grass" *(Imperata cylindrica)* to attract sambar deer *(Cervus unicolor)* and other large herbivores. There are patches of natural and anthropogenic grasslands and scrub areas in gaps caused by tree falls and landslides, in sediment-filled oxbow lakes, and in old infertile swidden fields throughout the otherwise forested interior (Puri 1996). Sago palms (e.g., *Arenga*, *Caryota*, and *Eugeissona* species) have been manipulated through pruning, weeding, and even transplanting. The harvesting of immature edible leaf buds of hill sago, *Eugeissiona utilis*, by Penan and other foragers may also have increased the size of stands of sago palms, known as *birai*, by stimulating vegetative reproduction. Continued harvesting may in fact perpetuate the *birai*, such that they come to dominate dry ridge tops for hundreds of meters at the expense of many other plant species (Brosius 1991; Puri 1997b).

Perspectives of hunter-gatherers as destructive, savage destroyers of nature are as extreme and fallacious as views of them as benign conservationists. The way forward is to transcend the debate and focus on the conditions under which forest collectors are least destructive and most productive (Dove and Sajise 1997). The Kayan Mentarang area, which the Penan consider their home, is not particularly well studied, but it still strikes one as being less rich in species, especially animals, than typical lowland forest areas. The mountains are steep, the soils are not very fertile, and there is a long history of human habitation and disturbance, all factors that would suggest a degraded and depauperate flora and fauna. Yet the dynamism of both human and natural disturbances interacting across the area and over time seems to have created a diverse biocultural landscape with high habitat and species richness, at least in those domains that have been investigated to date (Dove and Nugroho 1994; Wulfraat and Samsu 2000).[9] The ruggedness of the landscape has restricted settlements to flat mountain terraces, ridge tops, and the banks of the larger rivers, leaving areas of steep land untouched and covered with old growth forest. Furthermore, it seems likely that the low human population density, coupled with a high rate of migration due to resource depletion, disease, and warfare, has helped lessen the impact of people in any one location. The constant migration of people has allowed time for regeneration of the forest as well as the colonization and use of anthropogenic habitats by wildlife. Even wild animals normally found in primary forest have been observed in abandoned villages, usually feeding in the fruit trees.

On the whole and over the long term, the Penan have trod lightly on their land. They have caused changes, however, to the distribution of fruit trees, sago palms, and grasslands and have retarded natural forest succession through burning, pruning, felling, and reoccupation of old campsites. The changes in settlement patterns and subsistence strategies that have occurred since the mid-1960s, including the adoption of swidden cultivation, manioc gardening, and repeated occupation of nearby fruit groves, have resulted in a shift in the locus and intensity of their impact on the landscape. Once extensive and distributed across a variety of natural and anthropogenic vegetation types, the impact of the Penan has become intensely focused on the nearby anthropogenic vegetation of the villages. One benefit of this shift has been a reduction in the reoccupation of old mountain campsites, which have high tree species diversity and in some cases a regenerating undergrowth rich in fruit saplings. It remains to be seen whether the influence of the Penan was indeed strong enough that left on their own, without recurring manipulation by visitors, this next generation of trees will survive and eventually dominate the canopy in former campsites. If that should happen, it probably will not be too long before the descendants of the Penan find their way back to the sites to harvest this fruit from their ancestors.

Acknowledgments

The research was conducted as part of the Conditions of Biodiversity Maintenance Project at the Program on Environment, East-West Center, and was funded in part by a MacArthur Foundation grant to Dr. Michael Dove and Dr. Percy Sajise. I thank my Penan Benalui hosts and friends, the late Billa Asang, Nawan Ilung, the late Usit Kojang, Siang Jalung, Kajan Kila, and the late Awing Kojang, for sharing their stories and history. I also thank Bisa Merang, Sapu Billa, Lalung Bisa, and Lukas Bisa for their enthusiasm and hard work in tracking down old campsites and conducting the field study. Thanks also go to Project Kayan Mentarang, WWF-IP, for logistical assistance in the field. Dr. Javier Caballero provided guidance on the use of NTSYS statistical software and data analysis. Drs. Michael Dove and Deanna Donovan provided comments on early drafts. I alone am responsible for the final work.

Notes

1 In addition to examining Penan forest camps, similar research is being conducted in current and abandoned villages of Kenyah swidden farmers in the Lurah. Preliminary results from this study will be referred to throughout this analysis, but the focus is predominantly on the Penan.

2 See works by Meilleur (1994), Balée (1989), Dove and Sajise (1997), Dove and Nugroho (1994), and Sponsel (1997a).

3 Some of this research has already been carried out for hunting (Puri 1992, 1996, 1997a), sago palm processing (Puri 1997b), collection of *gaharu* (Momberg, Puri, and Jessup 2000; Donovan and Puri 1998), rattan gardens (Sirait 1997), and rice swiddens (Syahirsyah 1997).

4 In the two years since the Penan left the village, many of the trails leading to these abandoned sites have become overgrown from lack of use. Apparently, the Kenyah residents seldom travel as widely as did the Penan.

5 Site L is adjacent to a temporary Merap village site, probably established in the late-1800s when the Merap were being chased by the warring Kenyah Uma' Alim out of the Lurah toward the Malinau river valley. The Penan may have chosen to locate their camp nearby because of the secondary forest and/or the presence of fruit trees, the result of the prior occupation of the Merap.

6 Of the commonly encountered fruit trees, those from the family Moraceae (e.g., *Ficus* and *Artocarpus*, species of figs and jackfruit or wild breadfruit) show tree rings, but the cause of these rings and their relationship to some temporal sequence is unknown. They may indicate times of slow or no growth caused by periodic reductions in rainfall, droughts, or even large fruit sets. Coster (1927), for East Java, lists eight species that gave exact results in calculating age from tree rings: *Cassia fistula, Pterocarpus indicus, Toona sureni, Melia azedarach, Homalium tomentosum, Lagerstroemia speciosa, Tectona grandis,* and *Peronema canescens.* Presumably, these and other species can be investigated for their accuracy in dating forests in more aseasonal East Kalimantan.

7 There have been important historical studies in human ecology (Alvard 1993; Brosius 1991; Dwyer and Minnegal 1991; Kartawinata et al. 1981; Moran 1993; Vayda, Jessup, and Mackie 1985), agroforestry (Alcorn 1984; Linares 1976; Padoch and Peters 1993), ethnobiology (Meilleur 1994; Posey 1985), historical ecology (Balée 1989, 1997; Posey and Balée 1989; Sponsel 1997b), archaeology (Roosevelt et al. 1996), and geography (Brookfield, Potter, and Byron 1996).

8 See Dove and Nugroho 1994; Gomez-Pompa and Kaus 1992; Jong 1995; Meilleur 1994; Moran 1993; Posey 1985; Saberwal 1996, 1997; and Western 1989.

9 Studies on biodiversity in Kayan Mentarang include McDonald et al. 1992; Engstrom 1993; Balen 1995; Puri 1997a; Sorensen and Morris 1997; and Wulfraat and Samsu 2000.

CHAPTER ONE APPENDIX

Table 1.1A *Migration history of a Penan Benalui family in Long Peliran since the 1950s*

Location	Date	Length of Occupation	Comments	Forest Type[a]	Plot
Upper S. Aran Valley/ S. Malinau	1950s	? years	Clients of Kenyah Uma' Alim (Lg. Metap)		
Lt. Iyu (S. Bahau nr. Lg. Peliran)	1950s–1960s	2 years	5 families move to Lg. Peliran area (BA, DK, JK, LA, UK); clients of Kenyah Uma' Badeng.	1F	H
Lt. Bollo (S. Bahau)	1960s	3 months	Flee disease; Long Lt. Bollo, close to village	L	
Ridge btw. Lt Bollo and Lt. Tekojang (S. Bahau)	1960s	6 months	Flee disease; Mtn. ridge S. of Lg. Peliran village	1F	G
Apau Tekojang (S. Bahau)	1960s	6 months	Sago processing site; next to *Eugeissona utilis* grove	1F	F
Lepu'un pet anak bala (S. Kedayan)	1960s	6 months	Flee disease; mtn. ridge between S. Ingan and S. Kedayan Kanan	1F	J
Lt. Jelaeban (S. Kedayan)	1960s	3 months	Fruit season camp	1F	I
Modung Nakan (Kedayan)	1960s	3 months	Fruit season camp	1F	M
Lt. Penan (S. Peliran)	1960s	2 years	Settlement in or next to old swidden fallow and planted fruit trees	Old 2F	E
S. Aran	1963–5	< 2 years	BA and JK families flee Javanese troops (SB born)		
S. Salo (S. Bahau)	1965	6 months	Return to Lg. Peliran area; old swidden garden	Old 2F	C
Lg. Lt. Iyu (S. Bahau)	1966	? months	Residence		
Lg. Peliran village		? months	Residence		
Lg. S. Diam Nau (S. Bahau)		2 years	Residence next to Kenyah Uma' Badeng swidden field	L	R

(continued)

Table 1.1A *(continued)*

Location	Date	Length of Occupation	Comments	Forest Type[a]	Plot
Lg. Peliran		? months			
Upper S. Diam Nau		3 months	Residence; fruit season camp	1F	Q
Lg. Peliran		? months			
Lt. Selaleng (S. Peliran)					
Lg. Tebelum (Lg. Lurah)	1974–75	1 year	Residence and swidden site	L	
Lasan Tada (Lg. Lurah)	1974	4 months	Fruit season camp of SB's family	Old 2F	T
Lt. Mejalin (Bahau)	1975	2 month	Fruit season camp of BM's family	2F	A
Lg. Belaung (S. Kedayan)	1975	? months	Fruit season camp and residence of SB's family	2F	
Lg. Tebelum (Lg. Lurah)	1975–76	1 year	Residence and swidden site	L	
Lasan Tada (Lg. Lurah)	1975	? months	Fruit season camp	Old 2F	T
Lt. Belerung (S. Peliran)	1976	? months	Residence, rattan collecting camp (Edin born)	2F	
Lg. Tehgalo I (S. Peliran)			Upriver of Lg. Tehgalo II		
Lt. Poe (S. Peliran)					
Lg. Peliran		? months			
Lt. Belerang (S. Peliran)	1982	2 years	(ET born)		
Lg. Peliran		? months			
Lg. Tehgalo II				LG	B

(continued)

Table 1.1A *(continued)*

Location	Date	Length of Occupation	Comments	Forest Type[a]	Plot
Lg. Peliran		? months			
Lg. Tehgalo III			Across river from Lg. Tehgalo II	2F	
S. Batu Bala (Lg. Uli)			Lg. Uli school built; DK stays in Lg. Uli		
Lt. Sepan (S. Bahau)					
Lg. Peliran					
S. Kedayan (S. Lurah)	1985	4 years	Old Kenyah longhouse site; orchard; swiddens on Kedayan River (Across river from Plot O.)	VG	
S. Bollo Latung (S. Lurah)	1988	2 years	Swidden site	L	P
Lt. Beta (S. Kedayan)	8–9/1990	2 months	Fruit season camp (Mast)	1F	M
Cabang S. Kedayan	10/1990	1 month	Fruit season camp (Mast)	2F	N
S. Bollo Latung	11/1990	1 month	Swidden site	L	P
Lg. Peliran	12/1990	2 weeks	Holiday season in village	V	
Lt. Parot (S. Peliran)	1/1991	3 months	Rattan collecting expedition (BA, BM, SB, LU, TL)	2F	
Lg. Peliran	4–5/1991	2 months	In Kenyah house	V	
Hilir Lg. Peliran	6/1991	1 year	Swidden site (BA, SB, LU; AK dies)	2F	
Lg. Peliran	6–9/1992	4 months	In village, second year swidden; (BA dies; TL and family move to Binai)	VG	

(continued)

Table 1.1A *(continued)*

Location	Date	Length of Occupation	Comments	Forest Type[a]	Plot
Lg. Tehgalo II	9–10/1992	2 months	Fruit season camp (Mast) (BM, SB, LU, NI)	LG	B
Lg. Peliran	11/1992-3	2 years	Swiddens on the Bahau River (BM, SB, LU, NI)	VG	S
Lio Moduh (S. Bahau)	5/1994	1 year	Swidden site (SB, LU, NI)	2F	
Lg. Tehgalo II	1995	1 month	Fruit season camp (Sept. or Oct.)	LG	B
Lt. Boda (S. Peliran)	1995	1 year	Swidden site (UK dies)	Old 2F	
Lepu'un Kadeng (Lg. Pujungan)	3/1996	9 months	Leaves Lg. Peliran for Kenyah Uma' Lasan of Pujungan; swidden site	L	

[a]Forest type at time of occupation: 1F = mature forest, 2F = secondary forest, L = rice swidden, LG = swidden garden or orchard, V = village, VG = village garden or orchard.

Table 1.2A *Anthropological characteristics of surveyed Penan campsites and Kenyah Badeng orchards in Long Peliran*

Plot	Primary Activity ACT	Population Size (families) POP	Number of Occupations OCC	Duration of Occupation (yrs) TIM	Age since Abandonment (yrs before 1996) ABN
A	Fruit camp	2	1	0.25	21
B	Fruit camp	4	4	1.5	4
C	Fruit camp	1	3	1	1
D	Fruit orchard	—	—	—	—
E	Residence	5	1	2	30
F	Sago camp	2	1	0.5	28
G	Residence	5	1	0.5	28
H	Residence	5	2	1	29
I	Fruit camp	2	2	0.5	11
J	Residence	5	1	1	27
K	Fruit camp	1	1	0.25	6
L	Residence	5	2	2	22
M	Fruit camp	5	1	0.25	25
N	Fruit camp	4	3	1	6
O	Fruit orchard	—	—	—	—
P	Swidden camp	5	1	2.25	7
Q	Fruit camp	5	1	0.25	25
R	Residence	6	2	2	26
S	Residence[a]	4	1	2	3
T	Residence	6	1	0.5	21

[a]Abandoned Penan camp in Kenyah village gardens and orchard.

Table 1.3A *Biophysical characteristics of surveyed Penan campsites and Kenyah Badeng orchards in Long Peliran*

	Vegetation in and around plot[a]	Reported Age of Tree Vegetation (yrs before 1996)	Altitude (m asl)	Distance to Village[b] (m)	Topography of Site
Plot	VEG	AGE	ALT	DIS	TOP
A	Old secondary (20–40 yrs old)	21	420	541	Ridge, trail
B	Secondary (10–20 yrs old)	25	340	2,619	Streamside, trail
C	Old secondary (20–40 yrs old)	31	385	1,666	Hillside, trail
D	Kenyah orchards	54	260	59	Riverine, trail
E	Old secondary (20–40 yrs old)	30	530	1,666	Ridge
F	Primary forest[c]	28	560	1,847	Ridge
G	Mature forest	28	560	1,307	Ridge, trail
H	Mature forest	32	450	1,092	Ridge, trail
I	Old secondary (20–40 yrs old)	27	400	4,329	Streamside
J	Primary forest[c]	27	650	4,367	Ridge, trail
K	Mature forest	6	655	3,398	Ridge, trail
L	Primary forest[c]	22	745	5,559	Ridge, trail
M	Primary forest[c]	25	710	5,259	Ridge, trail
N	Secondary (10–20 yrs old)	6	380	4,119	Streamside
O	Kenyah orchards	48	310	1,506	Riverine
P	Young secondary (5–10 yrs old)	7	360	3,643	Streamside
Q	Mature forest	25	405	3,165	Ridge, trail
R	Secondary (10–20 yrs old)	26	295	2,645	Hillside
S	Village orchard[d]	54	280	0	Riverine, trail
T	Secondary (10–20 yrs old)	21	325	828	Ridge, trail

[a]Present day vegetation types translated from Penan Benalui ethnobiological classification.
[b]Distance is direct line, "as the crow flies," from site to center of Lg. Peliran village (plot S).
[c]Believed, by informants, to have never been cut for swidden agriculture.
[d]Abandoned Penan camp in village gardens and orchard.

2

Uneasy Bedfellows? Contrastin[illegible] Conservation in Peninsular Malaysia

Lye Tuck-Po

THIS CHAPTER JUXTAPOSES Batek environmental ideology with that embedded in the scientific mode of protected areas governance in Malaysia.[1] I will suggest that the local model of environmental relations has global relevance and can help reveal weaknesses in broader (statecentric) claims to biodiversity conservation. One underlying issue will become clear: there is no comfortable fit between scientific and local definitions of conservation. For example, we do not normally recognize as "conservationist" statements such as the following: if there were no people in the forest, the world would collapse; and it is forest peoples, and not urban peoples, who *jagaʔ həp* (guard the forest). These positions are commonly held among the Batek, who are mobile,[2] forest-dwelling, hunter-gatherers of Peninsular Malaysia. They call themselves *batɛk həp* (people of the forest). Elsewhere I develop the argument that this association of ideas is part of a broader expression of environmental stewardship (Lye 2002). The Batek certainly share the general conviction that the environment is degrading at an alarming pace and that there needs to be a radical shift in environmental values. Where they depart from the scientific model is in how they conceptualize and represent *people*—that it is necessary to have people in the forest, guarding it and the world from collapse. This is the theme of this chapter's discussion.

With the Batek, the forest is never defined as a static and distanced space, divorced from the everyday concerns of life. The continued existence of the forest, *həp*, depends on the stability of the *teʔ*, defined variously as the ground beneath the feet, the soil or earth (the physical stuff) of which it is made, the land that keeps the underground waters from welling up and inundating the world, and in some limited contexts perhaps even territorial space. To hold up the world is to keep the *te?* from fragmenting, and this requires careful observance of a range of practices. The most straightforward of these is to ensure that there is a minimum of tree cover to prevent soil erosion. There is also a rich set of rituals and taboos whose overt objective is to prevent humans from angering the spiritual forces. Punishment for human misdeeds takes the stereotyped forms of thunder, rainstorms, and ground subsidence. Covertly, these beliefs are a form of social control and imply a sense that there is a natural order to the world that humans disrupt at their peril (Endicott 1979a). People are at the center of this cosmos; they have an important role in maintaining and reproducing forest dynamics and structure. What they do, or do not do, whether to the forest at large or to its numerous inhabitants—biodiversity in our terms—has recognizable implications for the forest. Batek representations put the emphasis on relational dynamics: those between people and those between society and the forest world.

Politically, Batek environmental knowledge has never been explored or treated seriously by protected areas managers and environmental planners. This oversight by planners may have actually helped to preserve the intrinsic fluidity and flexibility of the Batek knowledge system.[3] In contrast is the far more established knowledge of scientific conservation, which provides the basis for the official protection and preservation of biological diversity in Malaysia. While scientific conservation is not rooted in forest traditions, it is claimed to have universal relevance. I am not taking the extreme postmodern position that all knowledge is inherently so implicated in power plays and politics that only "texts" have any phenomenological reality (critiqued in Soulé and Lease 1995). What I recognize as problematic is the all out determination in official circles to replace "traditional" knowledges like the Batek's with their scientific counterparts, thus overlooking the potential of those knowledges to inform and illuminate problems of environmental management. I take the position that the Batek have something compelling to teach about biodiversity maintenance, but I also argue that their knowl-

edge is not disengaged from, nor can it be studied as unaffected by, the broader social and political milieu. What constitutes "tradition" is inherently problematic (Lye 2002).

One key difference between Batek and scientific models will emerge in this discussion: that the former is a profoundly anthropocentric vision (as discussed above) while the latter is ecocentric and therefore only uneasily recognizes the role of people and societal relations in shaping the environment. The key thesis of this chapter is that while the Batek privilege the presence of people science privileges the absence of people. For the Batek, conservation begins at the level of practical involvement with ongoing environmental processes. Conversely, for scientists conservation begins by taking people out of the environment. The scientific approach is most visible in the common practice of setting up wildlife reserves and habitat protection zones where human activity is restricted, circumscribed, or simply forbidden. As such, the "ethnographic space" of this chapter is Taman Negara National Park, which is the largest unbroken tract of forest remaining in Batek territory.

This chapter does not attempt to say whether local or scientific paradigms have a better grip on "reality." Different paradigms have different uses and values and should be evaluated on their own terms. But I do argue that successful conservation needs to be an accommodation of scientific and local knowledges. The argument develops as follows. In order to make both ends of the spectrum clearer, I draw a "line" between Batek and scientific conservation, that is, I typify them. In showing how they diverge on key concepts and how they merge in the everyday lives of the Batek, I intend to indicate the strengths or limitations of scientific conservation for this context. In so doing, I argue that biodiversity conservation needs to move beyond its technoenvironmental imperative toward the ontology of conservation and that biodiversity conservation will succeed by targeting not only the problems of nature, but also those of society.

Ethnographic Overview and Methodological Approach

The Batek are one of the score or so indigenous ethnic minorities of Peninsular Malaysia, the Orang Asli (Malay: "original people"). They live in lowland tropical forests (roughly 500 to 600 m asl) in Pahang, Kelantan, and Terengganu states. My studies have been limited to Pahang, where the population (as of mid-2001) was roughly 450 out of the estimated total of 700 to 900.[4]

In the post–World War II era, government policy on Orang Asli has been premised on the need for assimilation, by which is meant assimilation into (Muslim) Malay society (Carey 1970). This has led to a planned program of "spiritual development," meaning Islamization. Materially, state intervention usually begins by shaping the mode of food procurement, that is, by promoting cash crop agriculture, and this is inextricable from the policy not to encourage the continuance of mobile ways of life (Jabatan Hal-Ehwal Orang Asli Malaysia 1961).[5]

The economic trajectory in Malaysia has been driven by export-led growth, with industrialization being the desired objective. Growing urbanization, land conversion, infrastructure building, and export-based plantation development (now mainly oil palm) are among the major causes of forest loss and biodiversity decline (for a recent overview of key issues, see Sahabat Alam Malaysia 2001). Much of the forest is heavily fragmented. For Orang Asli, these changes translate into the loss of subsistence bases, sociopolitical impacts aside. Land security now ranks high on any list of the Orang Asli's pressing needs (Nicholas 2000). In Batek territory, large-scale logging and land transformation began in the early 1970s in Kelantan; Kirk Endicott (2000, 210) estimates that roughly two-thirds of the Batek area is now lost beyond regeneration. Half the people in Kelantan and most in Terengganu have become semi-sedentary cash crop farmers. Currently, the Pahang landscape ranges from advanced climax to recently logged forest. As long as forest cover remains, the Batek continue to regard logged areas as home and will return to them. Just over half of the Pahang Batek spend a majority of their time in the 4,343 sq km Taman Negara, the National Park that mostly sits astride Batek territory. The Batek are regarded as the original inhabitants of the park, but they do not have an administrative role and are not consulted on management issues.

Mobility remains a fundamental characteristic of Batek social life. Inter-group boundaries barely exist, if at all. Groups materialize in the setting of the camp (*hayaʔ*). These camp groups (the average size is 36.2 people, with a range of 6 to 70) are not political entities but rather associations of people who are united by kinship bonds, friendship, and shared interests; membership is open and fluid (Endicott and Endicott 1986). A camp group may average two weeks per encampment; the distance between successive camps is about two walking hours or roughly 9 to 10 kilometers. After three or four months, camp groups disband and splinter groups may disperse to other river valleys, joining and forming groups anew. On a daily basis, individuals might move in and out as social, economic, ecological, and political conditions change.

Economically, hunting and gathering have been the production base, but the Batek also undertake a variety of other activities. There are daily, seasonal, and annual changes in production activities. The core of the economy remains subsistence-based hunting and gathering. This does not mean that quantitatively more time is spent on such activities; rather these activities are undertaken even when there are competing income-generating opportunities. They have high cultural value. Seasonal activities include honey and fruit collecting. The main source of cash income is commercial extraction of forest products, primarily rattan (Batek: *ʔawey*, mainly *Calamus* spp.) and aromatic woods or *gaharu* (Batek: *baŋkol*, *Aquilaria* spp.). When opportunities arise, men may do some day laboring, and occasionally there is some casual planting of fast-growing vegetables (Endicott 1984; Lye 1997, 69–76). Primary reliance on agriculture, however, is the least favored of these activities. Those living close to the headquarters of Taman Negara National Park are also heavily involved in tourism, both in hosting the visits of tour groups to their camps and settlement and in guiding tourists to the summit of Gunong Tahan (the highest mountain on the peninsula).

A third of my fieldwork was spent inside Taman Negara. Because the relationship of the Batek to the practice of scientific conservation is more heightened there, I primarily refer to this experience. My understanding of Batek knowledge, on the other hand, is drawn from interviews and observations that occurred throughout the fieldwork. My analytical strategy, to disentangle the relationship between conservation and the Batek, was to see how the institutional perception of the Batek emerges in policy decisions and is made manifest in signs and symbols that in turn reveal how Taman Negara authorities think about the Batek and their world. Among these signs and symbols are such things as place names, poster displays, and directional markers. These are not as irrelevant to biodiversity conservation as they might seem. I argue that they are part of a communication system whose purpose is to "transform" (administratively, cognitively, and linguistically) the forest into a park that can be understood by the average visitor. What and who belongs in that place, what kind of behavior is appropriate and desirable, and how that place should be known and used are among the messages, implicit and explicit, that are "sent out" to visitors through these signs and symbols. In other words, the design of the park draws from a universe of ideas regarding the proper relationship between people and the forest. My argument is that these ideas are not just products of a scientific approach to conservation or a value-free approach to park management. They convey the authorities' sense of what the Taman Negara landscape should be like, and therefore there are implicit issues of power and politics. As such, these signs and symbols are cultural objects that reflect the worldview from which they come and can be subjected to cultural analysis.

The Batek Relationship with Science

History and Background of Taman Negara

Taman Negara was founded in 1938–1939 and is managed by the Department of Wildlife and National Parks (DWNP) of the Federal-level Ministry of Science, Technology, and the Environment. The DWNP's mandate is limited to Peninsular or West Malaysia. (The protected area system is different in the eastern Malaysian states of Sabah and Sarawak, and is not addressed here.) Tellingly, the DWNP's many functions do not include "the authority to prevent the conversion of wildlife habitats into agricultural or other development purposes" (Misliah and Othman 1996, 154).

Historically, the primary goal of Malaysian national parks was the protection of wildlife habitats. Taman Negara is the oldest of the parks. Forest reserves—for protection of the timber supply—were created on the Peninsula as far back as 1884 (Burkill 1971, 206; Potter 2003). By the 1920s, with increasing alienation of land for settlements, agriculture (smallholder and estate-based plantations), mining, hunting, and poaching, faunal populations were in a clear decline (Burkill 1971, 207). The conservation lobby was relatively small. Cant (1973) notes the following for Pahang.

> From 1923 onwards District Office files contain many letters requesting the destruction of elephants or tigers which were damaging crops and houses or attacking livestock and human beings. Invariably the Game Department[6] investigation involved a long administrative delay and invariably the conclusion reached by the Department was that the damage done was very small and the animals concerned were harmless creatures.
>
> (114 n. 16)

By the close of the decade, people-wildlife conflicts were probably becoming impossible to ignore.

A special Wild Life Commission, headed by T.R. Hubback, was given the tasks of studying existing regulations, verifying allegations of agricultural damage by wildlife, and proposing solutions—in short, how to combine conservation with the protection of lives, livelihoods, and the economic-agricultural sector. Hubback was also asked to investigate the possibility of excising land to set up what is now Taman Negara. His three-volume *Report* (Wild Life Commission of Malaya 1932) makes fascinating reading. From the many testimonies recorded throughout the country, it seems clear that animal behavioral patterns, including their territorial distributions, were changing dramatically as forest was being cut away and that the populace—colonial planters, Malay settlers and villagers, Chinese chambers of commerce, and aborigines—was sharply divided over the best course to take.

Hubback's recommendations, over and above the specific conservation details (to be managed by a centralized Game Department, the forerunner of the DWNP), included the establishment of "inviolable sanctuaries" for the wildlife. The then Gunong Tahan Game Reserve (first granted protection in 1925) should be expanded and its status raised, and this first national park—eventually named King George V National Park—should serve the role of wildlife sanctuary, breeding ground, and refuge. Interestingly, the *Report* does not seem to contain any first-person aboriginal voices (in contrast to the abundance of colonial and Malay voices); aborigines are mainly represented through the sympathetic observations of others. The Batek ("Pangan") appear as informants advising Hubback on a stretch of the Pahang-Kelantan border and are mentioned by a Tembeling Malay headman thus: "The Pangan merely wander about and do no harm to anyone or anything" (Wild Life Commission of Malaya 1932, 319). Right at the outset, then, indigenous zoological knowledge was not taken into account.

In the 1987 *Taman Negara Master Plan* the policy objectives were explicitly declared as being based upon the American National Parks Service model (DWNP 1987, i), which in turn privileges the wilderness concept (Cronon 1996). Official policy (which is not a statutory provision) permits the Batek to continue living in the park and to travel in and out at will, but it does not permit commercial harvesting of forest products, including fauna (DWNP 1987). Notably, Taman Negara has been characterized as "an area of second-rate agricultural and forestry potential" (Anon. 1971b, 113); only 15 percent of the land is considered suitable for agriculture, minerals are absent, commercial timber is poor, and so are the sedimentary-derived soils. Indeed, one reason why it was possible to alienate so much land for the park in the 1930s was the absence of clear commercial interests there.

The goal of the park, as stated in the enabling enactments, is "the propagation, protection, and preservation of the indigenous fauna and flora of Malaya and of the preservation of objects and places of aesthetic, historical or scientific interest" (DWNP 1987, 3). This biological diversity is "to be kept in trust for all mankind, with controlled access for scientific research, education and tourism" (Anon. 1971b, 113). In the *Master Plan*, the priorities, in addition to the maintenance and management of biodiversity, the ecosystem, and scientific research, include an increased emphasis on recreational development or tourism (113; DWNP et al. 1986, 4–6). This final priority is rooted in the belief that "Taman Negara is the heritage of all Malaysians none of whom should be denied the reasonable opportunity to visit their National Park" (DWNP 1987, 11). No provision is made for the prior territorial rights of the Batek.

Recreational development is a controversial issue, as DWNP officials believe that economic development must be balanced against the imperatives of biodiversity conservation (DWNP 1987). Numbers of tourists per annum have increased dramatically in the last twenty-five years. My calculations from 1980–96 visitor statistics show an average annual increase of over 18 percent for Malaysian tourists and over 17 percent for all tourists. The average number of visitors per day grew from 1.41 in 1969 to 139.41 in 1996 (visitor numbers released by the DWNP in October 1996; analysis mine). Undoubtedly, growing interest in the environment coupled with increasing affluence in Malaysian society are among the reasons for this increase. It is also a result of government policy, for since the late 1980s tourism has been recognized and promoted as a major source of foreign exchange. The perceived national importance of tourism came out most clearly in 1998, when the minister of science, technology and the environment announced that pollution indices tracking the impact of the regional haze on locations around the country would no longer be published in the newspapers because this might frighten off the tourists. To take advantage of and promote tourism at the park, a private company, the Taman Negara Resort, has run the major accommodation facilities there since the late 1980s.

With the boom in tourism, socioeconomic life has changed dramatically for those villagers living at the margins of the park. If we look at the visitor statistics again, on an average day in 1996 there were probably half as many tourists as there were Batek in the park—and this excludes DWNP staff, their families, and the Taman Negara Resort workers, who are all based mainly on the Kuala Tahan headquarters side along the Tembeling River. This suggests not only the threat of crowding in the park but the reshaping of local society: from "sleepy" backwater to a microcosm of world society. The Batek have the forest to themselves in the interior, off the main tourist trails. At Kuala Tahan, they are a distinct minority. A common

Table 2.1 *Average numbers of Taman Negara visitors per annum*

Period	Average numbers of visitors per annum
1971–75	1,264.2
1976–80	3,699.6
1981–85	8,010.4
1986–90	14,042.8
1991–95	30,975.4

argument is that tourism supports the broader goal of conservation because it brings more people into the forest and therefore improves general awareness of environmental issues. The DWNP's concerns that tourist pressure will have a deleterious impact on the forest (Yong 1992) are generally overruled.

Opening the Park, Shutting out the People

As the foregoing shows, the DWNP faces the common dilemma of having to balance conservation and economic imperatives. Conservation planning demands a long-term vision: preserving the park so that it is still available for future generations means that success is measured by how long and in what ecological state the park can continue to exist and how well it maintains biodiversity. The reality is that for conservation to remain in the governmental vision it must show fiscal payoffs. State directors of DWNP Misliah and Sahir (1996, 158), discussing general wildlife conservation problems, point to the "ignorance (wrong attitude) of the decision-makers of the government," who "perceive that wildlife or even nature conservation has nothing to do with human livelihood." The effects of this attitude include "the rejection of any project proposal concerning wildlife conservation. Proposals for the establishment of more national parks, state parks or wildlife reserves, for example, were rejected by their giving various reasons, no matter who made the proposal, whether NGOs [nongovernmental organizations] or government agencies themselves. Worse still, the budget proposals for implementing conservation programs and activities are always being slashed to the bottom line. The same fate happens to proposals for funding external staff." In such a regime, increasing tourist numbers in the national park is probably one of the most desired economic indicators of "success."

My problem lies in the very ideology underlying DWNP policies and indeed its very existence: scientific conservation. Conservation ideas have come a long way since 1932. Today the globally established categories of protected areas are enshrined in the World Conservation Union (IUCN) protected area management categories (see its Web site <http://iucn.org/themes/wcpa/iucncategories-definitions-english.pdf>). Fundamentally, these categories range along a continuum of exclusiveness, with the most inclusive categories being areas where various forms of human activity are permitted (Bates and Rudel 2000; Robin 2000). Taman Negara is somewhere in the middle. The common principle underlying all such areas is that human communities are a potentially disruptive force in the park ecosystem and historically the park administration has taken a "fines and fences" approach to keep people out (Hughes and Flintan 2001, 4). The premise, in Lohmann's words (1993, 203), is that nature is "an industrial-free and therefore ([conservationists] assume) human-free reserve" (see also Ghimire and Pimbert 1997). While this opinion can be held in theory, it is not so easy in practice.

Taman Negara's borders are vulnerable to commercial development and encroachment. On the Trengganu side, some parts of the park are flooded within the catchment area of the Kenyir hydroelectric dam, and new dams are in planning in Kelantan. The highway to Gua Musang (a town in Kelantan) slices through a corner of the western part in Pahang. Many of the fringe forests outside the park are in various logged conditions. The ecological ideal of maintaining unbroken stretches of forest corridor for the free passage of wildlife is probably elusive in some places. Everyday problems include poaching and encroachment. Politically, the DWNP has been under pressure to make the park more conducive to tourist access, and this it has done by establishing additional points of entry into the park to complement the traditional one at Kuala Tahan (where lies park headquarters) as well as by eliminating the closed season at the end of the year. But if the agency goes too far in this direction, it will compromise its own existence and the goals of environmental conservation.

The classical response has been to make a case for a holistic understanding of biodiversity. An article in the *Malayan Nature Journal* by a former director general of the DWNP states this position thus: "We try to maintain the Park as a well-balanced system of wildlife that will serve the purposes that are outstanding there—recreational, aesthetic and scientific" (cited in DWNP 1987, 3). But this is where scientific conservation's own ideology works against itself. Conservationists try to mobilize support for a value of biodiversity that is hard to demonstrate; that is, they try to make the abstract—Western—concept of biodiversity into a meaningful idiom that can be assimilated into everyday systems of values and practice. But in fencing off that biodiversity and containing it within boundaries they may make it spatially and conceptually *distant* from everyday social life. In Ingold's words, this method "is like putting a 'do not touch' notice in front of a museum exhibit: we can observe, but only from a distance, one that excludes direct participation or active 'hands-on' involvement" (1994, 10). The DWNP 's words of choice for the role of humans are evidenced in its adoption of this common slogan: "Take nothing but photographs. Leave nothing but footprints."

Not only does this approach send out contradictory messages, it alienates the park authorities from the reality they seek to administer. As rangers themselves recognize, the park is not an empty wilderness. It contains a wealth of ecological, archaeological, historical, and cultural features (Adi 1989). My conversations with park tourists reveal that they know little of this background; it is not part of the educational "package" they are exposed to upon entry into the park. The traditional Batek homeland once centered around the area from Kelantan down to the Kenyam Valley. The limestone pillars in the valley are among the central iconic features of the Batek landscape; they are considered *pəsakaʔ* (heritage). From the scientific point of view, "the area is very poorly known and no aspect has been studied in detail" (Cole 1985, 2); from the Batek point of view, enough is known to give the pillars sacred status.

Furthermore, the proliferation of Batek names for rivers, streams, and landforms elsewhere in the forest clearly shows an ongoing cultural relationship with these places. As Frake points out, "unlike persons, whose creation precedes naming, places come into being out of spaces by being named" (1996, 235). When locals have established a deeply felt relationship with a place, they do tend to give it a name, which implies an emotional investment in it, and this makes the place more than just a transit point or physical space—a home, in fact. The Batek place names I recorded often show narrative and mnemonic content, for example, *Tɔm Sɔʔ Gajah* (Elephant Carcass River), *Tɔm Pacẽw* (Water Monitor Lizard River), and *Wɛ̃c Sok Bawac* (Pig-Tailed Macaque Fur Walking Pass). All of these encode animal sightings and encounters: an elephant died here, a water monitor lizard was seen there, they found pig-tailed macaque fur there. Or the name may be that of a plant, such as a tree species: *Tɔm Mahaŋ* (Macaranga River); *Ləpan Dok* (Ipoh Flat)[7]; *Tɔm Tuk* (Wild Banana River). Even now, the names evolve and there may be more than one name for a place. For example, *Tɔm Pənaceʔ (pənaceʔ* is the nut *Pangium edule)* is also known as *Tɔm ʔeyBajaw*. Somewhere down the line, the river associated with *Pangium edule* trees became associated with one of *ʔeyBajaw*'s turtle-fishing expeditions or vice versa—it is not clear. *ʔeyBajaw* is a living person. It is not unreasonable to expect that name evolutions are always happening as conditions and observations change. This is one way in which local histories are coded in the landscape. It is, indeed, their place of origin.

Representing the Batek

There is little symbolic indication of the Batek's presence in Taman Negara. Most of the place names do not appear on the official maps. Just one campsite, as far as I know, is named for a Batek individual (Keladong Camp). Even when names are recorded, the significance is lost. Few if any tourists are told that the Blau River and the Blau Jetty in Taman Negara take their names from the Batek word for "blowpipe" *(bəlaw)*. Or that the Tahan Trail, so popular among trekkers to the summit of Gunong Tahan, is part of an ancient trade route with great significance for the prehistory and history of Orang Asli—that these mountain climbers are, for part of their route, walking in the footsteps of the ancient travelers (Benjamin 1997). Few of the numerous streams and creeks show up as named places on the map.

The park enactments make no provision for the Batek. We would be hard pressed to find many official documents about their place and status in Taman Negara.[8] The most definitive statement is expressed in the *Master Plan*: "The department has to be quite pragmatic in accepting the fact that a certain number of Orang Asli have always lived within [the park] borders and will continue to do so in the future" (DWNP 1987, 23). On the one hand, the DWNP practices an "inclusionary" model of park management, which is certainly better than banning all human interactions with the forest. But, the *Master Plan* continues:

> It should also be noted that the Taman Negara Enactments make no reference to the aboriginal population that naturally inhabit the Park. There is also a department that has jurisdiction over Aboriginal affairs[9] and it is their general policy to integrate these people into modern society. For these two reasons there can be no policy regarding the preservation of the ethnicity of the Batek in Taman Negara in this document.
>
> (24)

Certainly this official position does not amount to a ringing endorsement of cultural-environmental relations in the park. Ingold's comment seems entirely apropos here.

> The presence of indigenous hunter-gatherers in regions designated for conservation has proved acutely embarrassing for the conservationists. For there is no way in which native people can be accommodated within schemes of scientific conservation except as *parts of the wildlife*, that is as constituents of the nature that is to be preserved.
>
> (1994, 11; emphasis in the original)

This identification of the Batek with nature shows up in many representations of the park. Let us look at two cultural objects now. The first is the controversial image that the Taman Negara Resort has used to entice visitors. The photograph, which has appeared on posters, press folders, brochures, postcards, T-shirts, and the Internet, shows a young Batek boy standing on rocks in a river, shyly holding the ubiquitous blowpipe. (The image became controversial when the Batek realized that they were being used to "sell" the park and were not getting any compensation for it.) On the resort's Web site (<www.tamannegarareort.com>), the caption below the photograph reads: "Only 10 years old, Kupeng has learned the ways of the nomadic Batek tribe. Adept with the blowpipe, small game are his prey. The natives of the rainforest are shy, gentle folk. Their philosophy: 'take little from nature that nature alone cannot heal.'" The fact that the Batek do not have a word equivalent to nature—or that they are unlikely to have made this attributed statement—is not important to the framers of this caption because what matters is to sell the park and the park's purpose is to promote "nature."

The second object furthers the message. It is a poster display on the Batek way of life that is on exhibit at the Taman Negara Interpretation Center in Kuala Tahan, the headquarters of the park. A selection of the text follows.

> Where do they go from here? On the surface their traditional lives appear ideal—they are in harmony with the forest, taking nothing from it that cannot be replaced by Nature.
>
> But ...
>
> Forest clearance outside Park boundaries has restricted the territory available to them and the law restricts their activities within the Park ... .
>
> Choices are limited: to leave the forest means embracing the problems of the urban or rural poor, as the forest skills they possess help little in other environments.

These words aptly summarize the received orthodoxy. I would argue that this image and text articulate the central tension in the relationship of the people not with any real or potential life in "other environments" but to the environment in which they presently live. The display allows us a glimpse into how Batek culture and knowledge are treated by the park authorities.

It is at base a sympathetic portrait of the Batek way of life. I suspect that most Malaysians viewing this display would not find anything amiss because the assumptions about Orang Asli are taken for granted and rarely questioned publicly. The display sketches out the main foraging activities, focusing on the Batek use and understanding of natural materials. It skips over their the cultural and social life. Finally it bluntly states that there is a possibility that the Batek will be ejected from the park, though such a move would be destructive, for the modern world has no place for the Batek. To remove the people from the forest, the text implies, would lead to grave social consequences; to allow them to live in the forest, on the other hand, is to risk greater pressure on the park's natural resources. This ambivalence seems to reflect the contradictory ethos of the conservation project, both embracing and distancing the Batek's place in the park.

Two things merit brief attention. The first, as with the boy's photograph, is the use of human presence in the park to culturalize the forest. Because the larger project aims to instill awareness of the natural heritage, it suits the interests of park interpreters to promote the Batek's relationship with the environment: the Batek, says the display, take "nothing from [the forest] that cannot be replaced by Nature."[10] This is a classic example of the tendency to identify hunter-gatherers as people "allegedly lacking the capability to control and transform nature" (Ingold 1994, 10–11). By implication, this orientation toward the forest is preferable to the visitor's urban lifestyle; hence, a separation is drawn between the culture of the visitor and the nature of the Batek.

Let's turn now to the assumption that "the forest skills they possess help little in other environments." This implies that the forest is what gives Batek knowledge (as revealed in foraging skills) its relevance and makes it unique. By implication, these skills have no relevance elsewhere. This perception misses the point of what knowledge is. Food procurement demands the knowledge not just to do (practical or procedural knowledge) but to look, search, discriminate, rank, remember, hear, track, compare, connect, map, and listen, that is, everything that comes under the category of human cognition. In other words, foraging knowledge, like any knowledge anywhere, is built on a base of problem-solving and tactical skills, largely implicit and beyond verbalization, that are necessary for living and are widely applicable to life anywhere, not just in the forest. This display, in short, essentializes the most visible parts of Batek life without giving adequate attention to the complexities of that life.

My point here is that in their choice of words and photographs the park authorities betray some highly revealing assumptions about the Batek. The Batek and their skills are good to look at and nice to think about; they are what stands out in the park. But when park managers present the skills as so unique that they are not transferable to other environments they set up a sharp boundary between "us" and "them," with "us" being the urban world that the Batek are pressed to join and "them" being the forest world. As such, the Batek belong to the forest but not to the world that is managing the forest, that of the park managers. In essence, the park sets up a boundary between their own knowledge and the knowledge of the people.

Probably three or four successive superintendents have run the park since the display was first put up, and no attempt has been made to revise or edit it. During this period, at least two anthropologists—myself and my predecessor Christian Vogt—have conducted officially approved fieldwork in the park, and neither of us was approached to comment on the display or make suggestions for its improvement. I would say that these assumptions are continuous with the broader style of management. As mentioned earlier, since even before the founding of the park, the Batek have been perceived as fairly irrelevant to habitat conservation. In the *Master Plan*, the major roles that the authorities can see for the Batek are as information couriers, drawing the managers' attention to locations of wildlife, trails, and scenic places (DWNP 1987, 23).

In all the years of visiting Taman Negara, I have never heard the park managers discuss how to draw on Batek knowledge to improve the overall running of the park. This is a common tendency in Malaysia. Tuboh, Sipail, and Gosungkit (1999, 212–13) comment about Sabah's Park Enactment in 1984: "To date, there is no written parks policy relating to indigenous peoples, and indigenous peoples' rights to parks are ignored ... the tendencies involved in formulating [conservation and management] strategies often ignore ... the indigenous communities." As a specific example of how the authorities see no conservation role for the Batek, Kirk Endicott (1996, pers. com.) notes that as far back as the early 1980s he recommended to the Jabatan Hal-Ehwal Orang Asli (Department of Orange Asli Affairs, or JHEOA) that the Batek be trained as forest rangers and that another point of entry to the park be constructed closer to the Batek camps on the Kelantan side. While the new point of entry (at Kuala Koh, Kelantan) has since been opened, it is independent of Endicott's proposal and is driven mostly by the need to relieve tourist pressure on the Kuala Tahan side. No attempts have been made to train the Batek as forest rangers.

The Batek Model of Conservation

A Plea for Conservation

My argument is precisely that Batek skills do have enormous implications beyond the confines of their local context. Let us hear now from the Batek themselves.

> The forest holds the *teʔ* [land, earth] together. When you cut down the trees, the *teʔ* breaks up. The rivers flood their banks, the *teʔ* becomes soft and fissures, the *dəɲaʔ* [world] ends. Our souls live upon the trees. The forest is the *ʔurɛt* [veins and tendons] of our lives. [He holds up his arm, and points to the veins and tendons therein.] The *halaʔ ʔasal* [superhuman beings] they say: "the *kəlaŋes* [heart] of the earth they had made." The superhumans, they remember the earth; they *haʔip* [miss] it. We can have a meeting. We meet, then we can discuss what to do. Discuss, decide, then we go ahead and do. Let's not give up the world. Let's not lose it. We should know how much to eat, how much to keep. *Gob* [Malays, standing for the government] think of building roads, of laying down oil palm. They kill the world. Where would we live? So they kill *dəɲaʔ heh* [our—everyone's—world]. In the past, we lived healthy. Now no longer can we want to be healthy. So everyone lives by common rules. We [Batek] miss the times of peace. We remember those times; we miss. We show the way. Already the earth is all cut up. The soul of the *tɔm* [rivers] is blocked. It's important to understand the danger. The rivers can no longer flow; they flood their banks. The soil becomes *ləkɔit* [soft]. They open up channels elsewhere; that's where the earth fissures. It's like we [all] make a living, and we become rich, but we lose the world. We should know how to keep. It shouldn't be that we enrich ourselves and kill the world. Our lives become shortened when we're too greedy. When we make a living, we should value the soul of the world. But if they *[gob]* don't value the soul of the world, well ....

These are the words of a Batek shaman, spoken two nights before I left the field. It is slightly shortened from the original text, which appears fully elsewhere (Lye 2002). This shaman—call him Tebu—is a greatly respected member of the society and here speaks as a representative. He is worried about the long-term effects of forest clearance, as are other Batek individuals, who share Tebu's sorrow and bitterness at the evidence of degradation around them. Tebu wants people outside the forest to recognize the far-reaching effects of forest degradation. Direct observation has shown the Batek that environmental conditions have deteriorated. If this trajectory persists unabated, Tebu asserts, humanity can destroy the world. A challenging and (to some) perhaps renegade idea is that problems in the Batek forest can be linked to the future of the world, even those parts of the world that are not obviously linked ecologically, geographically, politically, or economically to the Batek forest. Tebu also has a practical solution in mind. He urges us, people outside the forest (symbolized by his use of the inclusive form of the pronoun for "we" or "us," *heh*), to think about the consequences of taking too much from the environment and to consult with them, the people in the forest, in planning for the future.

The overt aim of Tebu's statement is to impress upon those who would cut down the forest what the consequences of their actions are. The Batek understand that their model of conservation has not hitherto been recognized, hence the need to send this message now. Tebu says that they want to sit down and discuss conditions with outsiders ("we can have a meeting"). They want to plan the future with us. They feel that we must be ignorant. In a sense, the Batek are correct, for their premises and ours are founded on different views of the environment. As evidenced by the "fence around the park" ideology, the environment is placed out of culture; nature is at the periphery of our vision and our everyday social lives. This, indeed, helps to account for the popularity of the national park: "Most people visit Taman Negara to experience the excitement of being in a tropical rainforest and to relax *away from* the bustle, cares and concerns of the modern, mainly urban world" (DWNP et al. 1986, iii; emphasis added).

To pay heed to the Batek message, I begin by interpreting their definitions of the environment. I examine the images that appear in Tebu's message. Two characteristics stand out: the use of bodily images and the focus on life, health, and death—processes of generation and regeneration. To the Batek, it is not the minute particulars of the forest that are important; theirs is a holistic vision in which "The forest is the veins and tendons of our lives" and "the soul of the rivers is blocked" while the superhuman beings say "the heart of the earth they had made." Here the use of bodily metaphors is deliberate, part of the broader pedagogical intent of the message. The environment has the same needs for sustenance as humans. To the Batek, the flow of human and animal blood causes grave anxieties, anxieties that emerge in the ritual complex. Veins and tendons are carriers and containers of blood. Just as we would do what we could to stanch the draining of our blood, so we should do what we can to prevent the "blood" of the forest from leaking into infinity. And the rivers have "soul." If the soul of the people depends on the continuance of the forest, the soul of the rivers depends on its freedom to flow on its own course. "Block" is my translation from the Malay word that Tebu used, *səkat* (Malay: *sekat*).[11] It captures the sense of flow being stopped, perhaps rudely, by something physical. That obstruction is made of soil, particulate matter, ground litter; when the land is fragmented, the soil matter washes into the river, and the flow of the river is blocked. With sedimentation, flowing rivers become ever shallower. The water has nowhere to go; it spills over the banks and pushes through the earth, looking for alternate channels, making the ground ever softer, ever more fragmented, ever closer to final fissure. But we are causing destruction not just in the terrestrial and visible realm. The superhuman beings who created the world and keep it going see what is happening, and they *haʔip*, which is that inexpressible feeling of longing for something or someone that is absent.

The superhuman beings *haʔip* the fertile, abundant, and healthy world that they created for us to enjoy; the Batek *haʔip* the past in which such terrors and images of doom were absent. *Haʔip*, if not salved, can lead to danger. It brings on *taŋɲõl*, the state of pining away when we cannot rise above our emotions, enjoy life; we refuse sustenance, and nothing can pull us out of our malaise. If we reach this stage of *haʔip*, we can become insensible to what is happening around us. Along the same lines, the superhuman beings can become indifferent to the concerns of this world and ignore the shamans' calls for help. In other words, *haʔip* has the potential to plunge the sufferer into spiraling depression; the next stage is death. If the superhuman beings were to perish because they are *haʔip*, there would be no possibility of regenerating the world. The destruction of the world would be assured.[12]

Tebu is not saying—don't cut down any more trees. He does not advocate a total shutdown. It is not that these things—roads, estates, township expansion—are a priori bad. What is bad is the loss of proportion and the sense that there is no end to forest degradation. As he said, "We should know how much to eat, how much to keep." He does not advocate taking land out of the development process. The idea that some places should be preserved for future generations (the founding premise behind Taman Negara) while others should be exploited commercially implies Cartesian notions: that land is divisible, is a form of property, and can be subject to quantifiable values. The Batek do not recognize property ownership of land (Endicott 1988). For them, I suggest, the environment, the cosmos, and the world are the same. What happens in one place affects conditions in others; what happens in the forest affects what happens outside it and vice versa.

In a way, this approach is a radical one because it implies a fundamental shift in values. If land is not divisible into conservation and development zones, ultimately the distinctions between "local" and "global" and "nature" and "culture" make no sense in Batek terms. This would suggest that the political use of boundaries and borders to enclose nature protection zones is similarly problematic to the Batek. In fact, the DWNP itself has long known how difficult it is to protect the park from the development impetus. Indeed, a road is being built on the other side of the Tembeling River (which forms the boundary of much of the park's southern and eastern portions); this new development, in expanding access to the area, will certainly have an impact on wildlife ecology within the park (WWFM 1986, 3–5).[13] Thus, the "fence around the park" has quickly become a "road around the park." The Batek's integrated approach, in contrast, demands a total transformation in habits of thought and action.

The Batek do not have a word for nature. Nature is not a thing "out there," to be set apart from "culture"; by implication, all areas have cultural values. This does not mean that the Batek do not appreciate the biophysical differences between forested and nonforested areas; rather, they understand that to protect the forest requires changes in the nonforested world, that is, in perception and approach. Just as in talking about life, health, and death they are appealing to universal processes of growth and decay, in their view the problems that beset their environment beset ours. The park managers, as noted earlier, believe that the Batek's "forest skills ... help little in other environments." This marginalization is being challenged. Tebu says that they want to talk; that means they want their knowledge recognized and they want to participate. In essence, if we continue to relegate their knowledge to the periphery we will die along with them. They cannot conceive of a separation between their concerns and ours. As Tebu says, if we do not love the world then what can they do? But if we reject the dichotomy between their vision and ours, and take the political steps necessary to listen to them, they could have much to teach.

Guarding the Forest

To follow up on the claim to conservation, I turn now to the statements mentioned at the outset, that there should be people in the forest, guarding it from harm, that *batɛk həp* (people of the forest) are the ones who *jagaʔ həp* (look after the forest). What does it mean to *jagaʔ* (guard) the forest? One afternoon I stopped ʔeyGk for a few minutes to talk about this.

His first comment was that the *batɛk ʔasal* (original people; i.e., deceased elders) send dreams to people to teach them how to *jagaʔ həp*. Those people inform others. Thus, before all else there must be *knowledge*—as received in dream communications from the spiritual forces and as communicated from one person to another. If forests are opened up or transformed, then the Batek cannot *jagaʔ həp*. ʔeyGk gave the example of Batu Sembilan, a Malay village on the edge of the Kenong State Park, near which Batek subgroups often camp. The forest there, he said, has been "cut" so much that it is beyond guarding. However, they can still *jagaʔ* the *Kəcɨw* area, which contains mainly logged forest. This is the second criterion of guardianship: the quality of the forest. It must not be thinned beyond a certain threshold, and it must contain the original trees. This is a covert definition of the *həp* that implies some degree of objective assessment.

What are the ways to *jagaʔ həp*? The shamans, by using their ritual knowledge to oversee the well-being of the cosmos, receive and convey intimations of perturbation and guard the people from harm. The shamans' roles are important. Kayəʔ, ʔeyGk's uncle, even says that if there are no shamans *jagaʔ həp* is not possible. As for everyone else, ʔeyGk said they can *jagaʔ* the forest resources; for example, they can make sure they do not exhaust the supply of rattan and can chase away intruders such as poachers who threaten that supply of resources. More important, they must take care of the *kayuʔ ʔasal* (original trees) because, ʔeyGk outlines, these trees attract bees, which produce honey, which attracts birds, which propagate the fruit. So if there are no trees, the birds won't come and *teʔ neŋ ʔum tahan* (the land cannot hold up).

Central to the Batek definition of *forest* is the presence of the original trees; these may parallel the scientific concept of "keystone species." To the Batek these were the first trees created on earth, those from which, in a sense, all forests have grown. If there were a ranking of trees on a hierarchy of importance, these species would probably be placed at the top. ʔeyGk's remarks imply—though this needs to be confirmed more generally—that when original trees are gone, the forest is no longer *həp* and guarding it is meaningless: technically possible perhaps but spiritually meaningless. There is some variation in the names and identities of these trees based on the experiences of each person. TaʔJamal named the *gɨl (Koompassia excelsa,* the tallest species of legume in Southeast Asia), *kəmpɛs* (Malay: *kempas, Koompassia malaccensis),* and *tagan* (possibly the Malay *penaga).* Tebu's list was *taduʔ (Oncosperma horrida,* a palm), *təkɛl* (Malay: *kepong, Shorea sericea*), *kəpoŋ* (Malay: *meranti, Shorea* spp.), and *kəmɔyɛn (Styrax benzoin,* a resin-producer). ʔeyGk himself listed *gɨl, təkɛl, kəpoŋ,* and *təmiŋ* (not identified but unlikely to be the bamboo of the same name, *Schizostachyum jaculans* Holttum). Many people mention the *təkɛl* and others also add the *gɨl.* Kirk Endicott's list from Kelantan was *cəmcɔm (Calamus castaneus,* a palm), *gɨl, təkɛl,* and an unknown species, *cinhɛr* (1979a, 34).

Agreement is not possible at this time. But there are some common characteristics of the trees designated as original. Other than the palm species, size, suggesting age, is an important differentiating element (all the palms mentioned are woody, pith-producing ones; woodiness seems to be the criteria for anything in the category of tree). In the context of ʔeyGk's remarks, the most important feature about the original trees is that they should be involved in the production of honey and fruit. Although these trees have their saliency, they do not exclusively define the forest. They are associated with the production of fruits, and it is that whole ecological process, including the role of organisms in pollination, that is significant. Thus, it is also the ecosystem processes that are important to the Batek. In Batek guardianship, the original trees are singled out because so much else depends on their persistence and health. They are the indicators of ecosystem integrity and therefore must remain in that ecosystem, reproducing it and being reproduced in turn.

As for the claims to guardianship, naʔGk, the wife of ʔeyGk, expressed this in no uncertain terms: *"batɛk gən gɔs hat həp gən jagaʔ həp, ʔayaŋ batɛk banar."* ("The people who dwell in the forest are the ones who guard the forest, not the people who live in town.") This is an explicit linkage between living in the forest (the way of life) and the capacity to guard it. Tebu voiced this issue implicitly, making a claim for relevance and posing a challenge to established environmental knowledge. This is a claim I heard many other times, when the Batek bitterly criticized outsiders (particularly park managers and Malay neighbors) for dismissing Batek claims to expertise. They argue that they know the forest and how to use it because they live in it, namely, that knowledge of place is contingent on being *present* and *resident*. This knowledge is a way to assert belonging and authority and is constitutive of identity.

I should state categorically that this is *not* a claim for ethnic purity. ʔeyGk said, if a *gob* lives for a long time in the forest (I used myself as the example), she or he can also *jagaʔ həp*—through long residence, even someone from the antiforest world could develop the knowledge necessary to guard the forest. Knowledge is embedded in a certain system of values. Batek *gob* (Batek in other states who have converted to Islam), he said, don't *jagaʔ həp* because they follow *kəmajuan gob* (Malay [modes of] progress). I asked Kayəʔ the same question. If a Batek *gob* lives in the forest, he replied, they can *jagaʔ həp*. But they do not—they ride motorbikes, live in houses, and plant crops. They, as ʔeyGk put it, *ʔikut krajaʔan* (follow the government) and therefore participate in the large-scale clearing of forests. This he contrasted with the Batek's approach, which is to cut a little bit as needed (e.g., when setting up the camps) because they *sayɛŋ ɲawaʔ* (value the soul). To rephrase this in terms introduced earlier, to *ʔikut krajaʔan* is to kill the world or adopt antiforest ways, while to *sayɛŋ ɲawaʔ* is to guard the forest. One can guard the forest because one follows the forest-dwelling *way of life*, with all the cognitive, behavioral, and normative implications of that, rather than because one is or is not born a Batek.

Now, to link all of this to the problems of running a national park, the *jagaʔ həp* ideology suggests that if there are no people in the forest, it is not possible to obtain the knowledge necessary to guard it from harm. Perception by and large is based on and enforces physical intimacy between perceiver and environmental images; as long as the religious taboos and principles are observed and people take appropriate precautions against danger and surprise, there is no need for, or value in, standing distant, aloof, and disengaged from the forest. While walking in the forest, a moment's carelessness or absentmindedness can lead to serious consequences. At best, one may become temporarily disoriented and lose the way. At worst, one could be killed, by accident, by animals, and so on.

I would contend, then, that there is a similar discouragement of aloofness and distance in how people conceptualize their relations with the forest. But the forest is not always benign; spirits, ghosts, and wildlife predators threaten human life. And people cannot recognize such threats without a highly developed sensibility of environmental information. This is radically different from controlling the forest *from the outside*, which is seen so clearly in the exclusionary mode of protected areas management. In the latter, the higher up the administration one goes the greater the decision-making power. And because the DWNP is a federal agency the most important decision makers are based in the capital city, Kuala Lumpur, people who rarely if ever step inside the national park to obtain direct, intimate knowledge of it and determine the best way to run it.

Conservation as Ontology

There are important philosophical differences between Batek and statecentric models of biodiversity maintenance. The focus on knowledge in the preceding section is worth exploring more deeply. Although Taman Negara is now over sixty years old, no one has ever asked the Batek how they maintain biodiversity. In the structure of governance, authority over the future of biodiversity is vested in the relevant government agencies. This necessarily biases resource management toward officially approved forms of "expert knowledge"—and in Malaysia it is assumed that this knowledge should come primarily from research and academic institutions. Local peoples are generally viewed as being "not exposed enough," and so when they are "involved" they are mainly the "targets" of educational programs rather than the teachers, as suggested by Tebu. All of this has implications for obtaining the Batek's support for the conservation project.

The park administration is focused on management, which means that the Batek's behavior is seen through the lens of laws and policies rather than local idioms and philosophies. As viewed by the Batek, scientific knowledge—or, more appropriately, its manifestation in park administration—is inappropriate and intrusive. An important example, and probably the cause of the initial tension between the Batek and the DWNP officials, was when the former were asked to move their original settlement from Kuala Tahan park headquarters areas because the authorities considered them "dirty." Dirt, as Mary Douglas has pointed out, has great symbolic meaning all over the world, but (like weeds) it is really just "matter out of place" (1970). In the park, it is represented by trash—a manifestation of the tendency to keep the park as clean of human evidence as possible. Batek trash symbolizes them as intruders and is therefore considered insulting because, as they often say, they predate the park. (The Batek are an easy target. The DWNP is unable to control land clearance and economic activities on the other side the Tembeling River, including the "floating restaurants" that have emerged to cater to tourists, which release sewage, effluents, and soil—another kind of dirt—into the waters and disrupt aquatic life.)

The Batek living in Taman Negara frequently cast aspersions on the park authorities' requests and demands. Once, for example, the carcass of a barking deer was found by rangers, who warned the Batek that they should not hunt such animals. In the first place the warning was inappropriate because the Wildlife Protection Act of 1972 explicitly permits Orang Asli subsistence hunting. In the second place, it was the only deer that had been captured that whole year, for it is one of the peculiarities of Batek hunting that they do not systematically target the largest, most vulnerable fauna such as tigers, elephants, seladang, bears, and rhinoceros and they tend to situate their camps away from these animals' migration routes (Endicott 1979b). So the rangers' warning was an overreaction and did not lead to the logical follow-up: to determine which Batek harvesting practices promote biodiversity. From what the Batek can see of the DWNP's model of biodiversity conservation, it is all about restricting their freedom and is totally alien.

For one thing, their respective epistemological assumptions are different. Let us probe this problem. As mentioned earlier, biodiversity conservation in Batek perception is less about the natural world than it is about relations with the various components of the universe. These components, both visible and nonvisible to external observers, form a single social field. If there are no people in the forest, disaster follows. What people can do to maintain the forest depends, as ʔeyGk suggested, on how knowledgeable they are. Two kinds of knowledge are relevant here: knowledge received from superhuman beings and deceased elders and knowledge that emerges from interacting with each other and the environment. Knowledge implies social relations: those between living and nonliving persons (in turn implying, perhaps, the need for historical continuity), among people in the group, and between shaman and society. One must have knowledge about the forest in order to guard it well.

The Batek often say that if they give up their space and their identity the superhuman beings will not be able to reach them. This is the central theme in Batek cosmology. People's actions inside the forest serve as conduits of knowledge from the superhuman beings. From these beings knowledge comes to save the world, to stave off ritual danger, and to prevent ecological collapse. Maintaining biodiversity, therefore, results from pursuing good relations with these superhuman entities. This in turn entails heeding what the superhuman beings tell them about environmental conditions. This is the mechanism through which the Batek can learn that current environmental conditions are making the superhuman beings feel *haʔip*. To receive this knowledge, what we might artificially mark off as ritual knowledge, one has to be physically present in the forest. Out of the forest, knowledge, and thus culture, are lost.

As for the other kind of knowledge, knowledge in a more prosaic sense, it results from a person's growing experiences in the world. There is plenty of evidence that Batek knowledge fits the emerging paradigm of local knowledge as "a practical situated activity, constituted by a past, but changing, history of practices" (Hobart 1993, 17). All domains of knowledge are in some sense historical. In practice, every exercise of knowledge is a "performance," but all performances are generated from memory—of typical examples, habitual ways of doing things, ideal-type models, and so on—and therefore are the outcome of a long process of trial and experimentation going beyond the lifetimes of individual performers or agents. Knowledge also owes its origin to social memory (what is taught as part of language acquisition and what is understood to be the pedagogical content of lore, myth, and belief).

Earlier ʔeyGk began his remarks with the issue of knowledge. In order to *jagaʔ* resource supplies and prevent overharvesting, the Batek must be able to monitor conditions, from ecological quality to encroachment and land circumscription. Monitoring depends on communicated knowledge and works only when everybody keeps their eyes and ears open. When someone spots something, he or she is expected to tell others about it. And when something important or urgent happens in one place couriers are sent to alert other groups about it. The importance of communication can be seen in behavior: everyday Batek conversations tend to include some comment about what is happening in the forest—the people actively monitor such changes and developments, as the emergence of flower blossoms, the migration of bees, the growth stages of fruits and trailside vines, and the tracks and traces of animals (see Lye 1997, 98–140).

In short, Batek knowledge responds to never-ending changes in circumstances and conditions; it is responsive to ongoing conditions in the social and ecological worlds. While in a metaphysical sense a knowledgeable agent is situated in an ongoing relationship with the entities of the extranatural world, in a practical sense his or her knowledge is firmly embedded in daily activities and social life. In order to draw on these various sources of knowledge, a person must be bodily present. That is, there must be an efficient mechanism by means of which knowledge can be transmitted, some means by which people can get together, congregate and socialize, and give and receive knowledge. In Batek society, this demand for co-occurrence comes out clearly in the practice of *jok* (to move away). To *jok* is what mobile populations do: leave one place for another and seek friends, food, and companionship elsewhere. To this list, I would add knowledge. This kind of knowledge is all inclusive and is also integral to the ongoing constitution of personhood and identity. What this means is that knowledge also emerges from the body's continuous movements across the landscape. Movement is something that political-economic agents frequently lament about when considering the Batek. To them, mobility (or "nomadism") challenges development, while sedentariness enhances it. The most telling demonstration of this is the official policy not to encourage any form of mobility. But in the Batek view of things sedentariness does not enable knowledge seeking and identity construction and it threatens the well-being of the forest.

In short, the Batek are a small group of people claiming space for culture and knowledge and in the process doing something that in their minds and their actions maintains biodiversity. Biodiversity conservation, in short, is not contained within a delimited body of ideas like a codebook everyone can refer to. There is nothing in the Batek ideology that comes close to an identifiable *adat* (custom or tradition) of conservation. Rather, the Batek make the intellectual connection between their own actions, the responses of their original people and superhuman beings, and the effects of these actions on the ecology and their own lives and deaths. This linking of different domains of action enjoins upon the people a consciousness of caring—if you do not care, you risk your life. This is what I consider an ontology of conservation, that conservation only works if it is integrated into people's sense of self, being, and meaning.

Conclusion: Paradigms Lost And Regained

I accept the position that fundamentally there is no qualitative difference between local and scientific knowledge (Agrawal 1995, Kalland 2000). As Colin Scott points out of James Bay Cree hunting knowledge: "If one means by science a social activity that draws deductive inferences from first premises, that these inferences are deliberately and systematically verified in relation to experience, and that models of the world are reflexively adjusted to conform to observed regularities in the course of events, then, yes, Cree hunters practice science—as surely all human societies do" (1996, 69). What has concerned me here are the philosophical bases of local and scientific models of conservation and this means that, in order to see where the emphases fall in these respective traditions, the argument has proceeded by typification.

Viewed from the perspective of scientific conservation, local models of conservation are valuable if they can be proven to enhance biodiversity (Meilleur 1994). There is a strong materialist bias toward reproducible, replicable models. These have their values, and one can certainly uncover Batek practices that positively affect forest growth and composition (Lye 2002). They have not concerned me in this chapter. What we are interested in is ideology and how it is reproduced in Batek society. Social scientists may argue about the degree to which ideas affect and determine behavior or whether ideas or behavior should be given causal priority in our investigations. Given the centrality of knowledge to the Batek's social lives, I see no reason why their behavior needs to be disengaged from their intellectual apprehension of life.

With this chapter, then, I wish to make a case for a broader understanding of biodiversity, of the knowledge needed to conserve it, and indeed of conservation itself. The intent of this set of studies after all is to identify mechanisms by means of which biodiversity is being maintained by local populations. I understand this to mean that we want to understand how biodiversity is perceived and treated by the people, not how they fit other people's notions. As demonstrated above, the latter tendency is a form of myopia, which promotes conflict between administrators and people rather than cooperation and collaboration. Thus, the challenge ahead is to create not the technological tools that will the better help us to "manage," "control," or "organize" biodiversity conservation but to create or *re*-create the social conditions that promote a more inclusive view of "nature"—not out of society, contained in parks and reserves, but *inside* it, integral to it.

The Batek insist that their mobile, forest-dwelling life is preferable to the life that outsiders enjoy. The Batek do not claim that life in the forest is easy. But they say that it is preferable to continue this difficult life and live within modest means than to give up their communications with the superhuman beings. Clearly, then, the knowledge that they gain, and the security and well-being enjoined by this constant interaction with the superhuman beings and each other, are central to their own existential dilemmas. This kind of approach is more difficult to reproduce in other settings, and as long as it is dismissed as "mumbo jumbo stuff" (in the words of one Malaysian government official, discussing indigenous religions more generally) (Dentan 2002, 156) it will not be. Which paradigm is most suitable for this complex environment? The answer goes to the heart of scientific politics (Lewontin 2002). Now there is the possibility of embracing and incorporating the premises and methods of alternate models. We have the Batek and, by extension, other indigenous models from around the world. The challenge here is not to demand that these alternative models conform to the technoscientific model but to consider them as ascendant ideas that inform and point to weaknesses in standard scientific premises.

In the same spirit, I do not believe that I can do justice to the Batek paradigm by bringing along empirical evidence alone. It seems to me that we also miss the point if we cast either local or scientific paradigms away and reject the integrative holism of the Batek vision. What we seek, especially on this thorny issue, is multivocal knowledge that will, in Dwyer's words, "contribute to our imagined understanding" (1996, 181). This will not be easy to do, especially given that the scientific paradigm—much as it may struggle for governmental support—is in the politically dominant position and decides most of the premises upon which local models should be assessed. It is worth trying. As political-economic conditions in the country change, as bureaucratic actors enter and exit the theater of conservation, an empty forest becomes a liability. The notion and praxis of conservation do not seem to outlast the exigencies of a historical moment. The Batek paradigm, on the other hand, has historical continuity and staying power, and for these reasons it is worth serious consideration.

Acknowledgments

Fieldwork was funded by the Wenner-Gren Foundation for Anthropological Research and the East-West Center, with additional support from the John D. and Catherine T. MacArthur Foundation. I thank the Economic Planning Unit in the Prime Minister's Department, Kuala Lumpur, for permitting me to conduct fieldwork in Malaysia and the Department of Wildlife and National Parks for granting me access to Taman Negara. I thank Michael Dove for useful comments on earlier drafts of this chapter and for making it possible for me to participate in the MacArthur-funded project. My thanks also go to Tim Ingold and Kirk Endicott for comments on earlier versions of these ideas. Finally, this chapter is but a small measure of thanks to the Batek.

Notes

1 To avoid cluttering the article with neologisms, I keep to the conventional distinction between "local" and "scientific" or "global." This is not to say that local knowledge does not have scientific merit or that scientific knowledge is free of folk assumptions and premises.

2 The more popular term is nomadic. Because of the derogatory connotations of nomadism in governmental discourse, I prefer to use mobile. Nomadic in the popular sense is usually associated with free-roaming, aimless wandering, animalian behavior, primitivism, and the like. (See, e.g., Brosius 1986 and J. Scott 1998.)

3 For an example of how external attention may shift the emphases of local knowledge, see Brosius 1997.

4 Most of my data was gathered over a period of eighteen months in 1993, 1995, and 1996; this is supplemented by additional insights from 1998, 1999, and 2001 field visits.

5 Coerced sedentarization is not part of this policy. The usual practice with the Batek of Pahang seems to be "gentle persuasion" as to the benefits of a sedentary life.

6 Until wildlife conservation was integrated under a single federal-level department (see below), responsibility for game control and management rested with state departments.

7 This translation is provisional.

8 I have been criticized for not producing more written documentation of the agency's perception of the Batek; this is the reason.

9 The reference is to the Jabatan Hal-Ehwal Orang Asli, the JHEOA.

10 This line should be credited to Ken Rubeli (1986), a photographer who worked for the DWNP as a ranger and park interpreter. The line quoted here comes directly from his book, although the interpretation center makes no reference to it.

11 To my knowledge, there is no equivalent Batek word.

12 Comparable visions are being offered by other Malaysian peoples, the Temuan and Semai (Balasegaram 1996).

13 This road was originally a logging access road. Improvement began in the late 1980s and early 1990s. Work was stopped by the Asian fiscal crisis of 1997.

CHAPTER TWO APPENDIX

Table 2.1A Glossary of Batek words

ʔawey,	rattan
batɛk həp	people of the forest
ʔɔrɛt	veins and tendons of the world
dəɲaʔ	world's end
həp	forest
jok	to move away
kəlaŋis	heart
tek	land
tɔm	rivers
halaʔ)	shamans
halbəw	paths
haʔip	miss; long for
hayaʔ	camp
sayɛŋ	care for

Part II: Failures of State Conservation and Local Responses

3

Microlevel Implications for Macrolevel Policy: A Case Study of Conservation in the Upper Citarum River Basin, West Java

BUDHI GUNAWAN, PAMPANG PARIKESIT, AND OEKAN S. ABDOELLAH

RAPID ECOLOGICAL CHANGES in Java have placed many watersheds in critical condition. The Citarum river basin, where agricultural development is taking place at a rapid rate, especially in the upper part of the basin, is one of several in West Java identified as a critical watershed needing rehabilitation. At present, recurrent floods in the lowest part of the Citarum Basin appear to be related to the conversion of the natural forest to agricultural and industrial development in the upper and middle parts.[1] Our research indicates that the rate of environmental degradation is exacerbated by lack of a coordinated governmental effort to manage economic development and sustainable resource use in the river basin.

This chapter, which is based on a study of conservation and socioeconomic conditions in the upper Citarum river basin, discusses the discrepancy between macrolevel policy and microlevel resource use, discrepancies that have contributed to the current situation of unsustainable resource management. Two interrelated problems are examined. First, we demonstrate that macrolevel governmental policies in Bandung District (in the southern portion of the Citarum watershed) are in conflict with conditions of microlevel conservation. This is indicated by fact that government-supported agricultural and dairy-farming development is carried out with little regard for ecosystem diversity, the carrying capacity of the ecosystem for this development, or regional socioeconomic needs. Second, the lack of knowledge about conditions of microlevel conservation has resulted in various governmental departments launching conflicting programs and policies regarding the management of natural resources within the upper Citarum river basin. As a result governmental policies have had the unintended consequence of increasing ecological degradation in the region.

Ecology of the Upper Citarum River Basin

The Citarum watershed, situated in West Java, Indonesia, crosses seven different districts. The river runs approximately 350 km northward from Mount Wayang in the southern part of Bandung District to the Java Sea. The catchment area of 6,000 sq km is the largest in West Java. Three large reservoirs, which function to generate electricity, irrigate rice fields in lowland areas, control floods, and provide a venue for tourism, were developed by damming the Citarum River (see fig. 3.1).

The upper reaches of the Citarum river basin are hilly. As the river moves toward the Java Sea, the topography changes from undulating hills to plains (see fig. 3.2). The altitude of the study villages ranges from 650 to 1,500 m asl, while Mount Wayang (the headwaters of the Citarum River) rises to over 2,198 m asl. At this altitude, the Citarum watershed has great potential for upland agriculture in vegetable cash crops. In the last two decades, development in this sector, including the development of a dairy industry has been intensive and has provided substantial income for local people. But there have also been substantial ecological costs.

The Mount Wayang forest measures approximately 1,000 ha and is home to a significant fraction of Java's floral and fauna biodiversity. This is a protected forest for the maintenance of the Citarum River system's hydrological functions, and it is currently administered by a semiautonomous subunit of the Ministry of Forestry and Estates, called the State Forestry Corporation (SFC) or Perum Perhutani. Mount Wayang is one of just a handful of

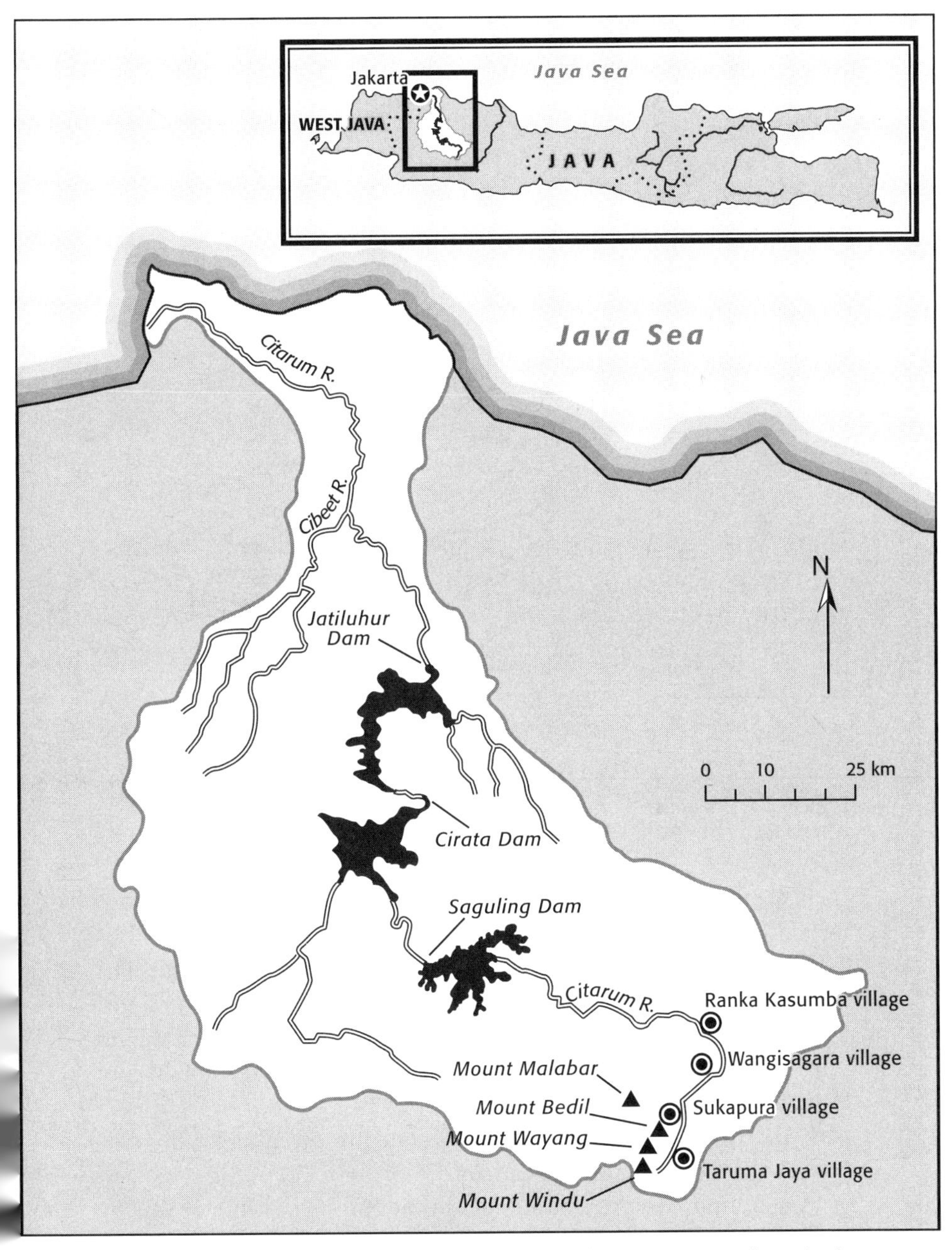

Figure 3.1 The island of Java and the location of Taruma Jaya village in the Citarum watershed

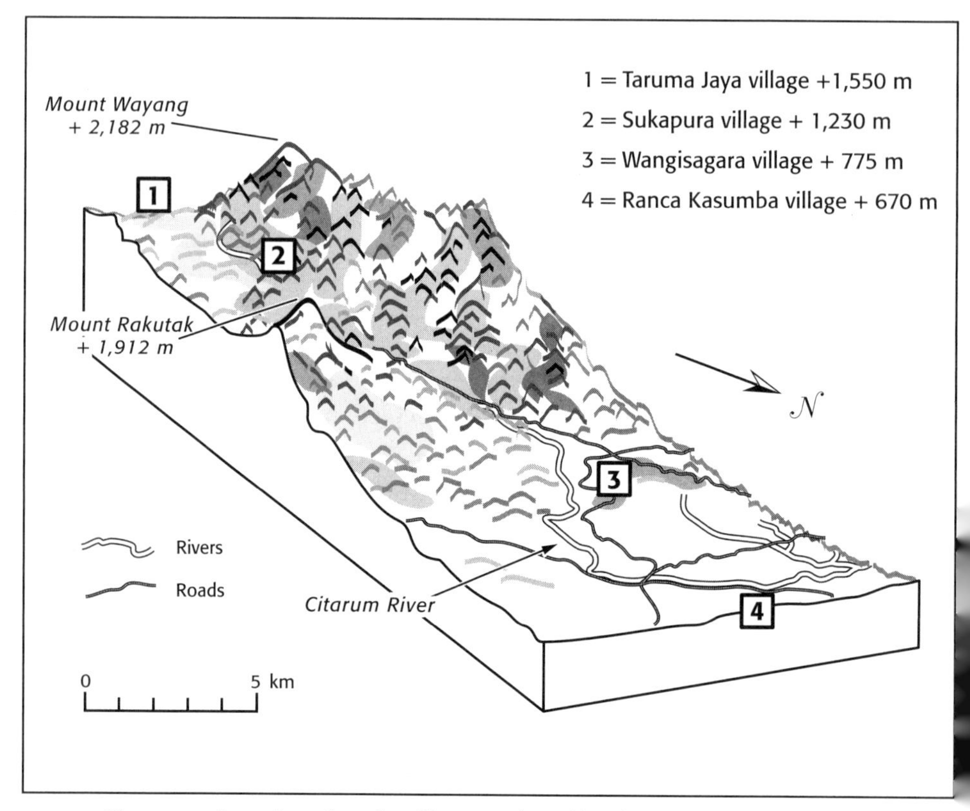

Figure 3.2 Location of study villages and profile of the upper Citarum river basin

remaining montane forests in Java, and it is currently under serious pressure due to rapid agricultural and industrial development. The mosaic of agroecosystems that surrounds this remnant of natural forest on Mount Wayang makes the forest look like a forest island within a sea of agriculture. Due to the limited extent of Mount Wayang forest, almost no part of it has been left entirely undisturbed. Although Mount Wayang is a state protected forest, it is also integral to the local agroecosystem.

Vegetation surveys undertaken in the forests of Mount Wayang indicate that, although the forest is under serious pressure, its floristic diversity is still relatively high (a total of 106 species, including 53 species of trees, were found in the sample plots). But further analysis, based on altitudinal differentiation, shows significant differences among zone I (1,660–1,830 m asl), zone II (1,830–2,000 m asl), and zone III (>2,000 m asl) in terms of the level of disturbance and the conditions of biodiversity. The middle zone, which is more disturbed than the other two, has fewer poles and saplings than in the peak and lower areas. In contrast, ground vegetation and pioneer trees are common in the middle zone but not in the other two areas. While there is a high degree of plant species diversity in the Mount Wayang forest, most of the species have low individual density. The scarcity of these species is due to their limited regenerative capacity; they are not able to compensate for the depletion of their populations caused by overexploitation.

A possible explanation of the difference between the zones involves the intensity of human disturbance, namely, illegal cutting occurs more frequently in the middle zone than in the other two. To avoid being caught by forest wardens, woodcutters (looking mostly for poles and saplings for construction material or fuelwood) tend to avoid cutting trees in the lower part of the forest (zone I). Instead, they go farther into the middle area (zone II) to cut trees, where they are more likely to escape the eye of the forest wardens. They do not go to zone III due to the more extreme topography, which is not conducive to the growth of suitable tree species (for a detailed analysis of the rates of disturbance in the three zones, see Parikesit and Yadikusumah 1997; and Gunawan, Abdoellah, and Parnawan 1997).

The altered vegetation pattern on Mount Wayang affects faunal diversity, particularly that of birds. Bird species diversity is generally high, and most of the birds found on Mount Wayang are mountain forest species, which are less tolerant of human disturbance. But our study found that bird species that are tolerant of human disturbance and commonly found in agricultural regions were dominant in the middle zone of the forest (e.g., the pearl-cheeked tree babbler *[Stachyris melanothorax]* and the oriental white-eye *[Zoosterop palpebrosus])*. Large populations of these two species in the middle zone indicate that there is a lot of brush cover and that human disturbance is relatively high. The blue nuthatch *(Sitta azurea)*, in contrast, is rarely found in the middle zone. It is abundant in the peak area of the forest where human disturbance is less intense and the forest is relatively undisturbed (Erawan et al. 1997).

The Agricultural Landscape of the Upper Citarum River Basin

The landscape of the upper Citarum river basin has undergone substantial changes since the first significant human settlement over two centuries ago. But the rate of change has increased markedly in recent years as agricultural development has intensified. An analysis of land use types and changes from 1900 to the present undertaken in five villages along the Citarum River is summarized in table 3.1. In general, we see a historic process of homogenization of the landscape (as indicated in fig. 3.2) with natural forests being converted to agricultural fields as subsistence agriculture has given way to both more intensive market-oriented upland agriculture and the development of the textile and brick industries.

Land Use Changes

Significant ecological changes in the upper Citarum river basin began in the nineteenth century when the Dutch colonial government logged and cleared natural forests for timber and plantations. This was followed by intensive human settlement and agricultural development. These ecological changes, particularly those since the

Table 3.1 *Dairy cattle population and growth*

	Number of Dairy Cattle		
Year	**Indonesia**	**West Java**	**District of Bandung**
1989	287,665	91,046	51,283
1990	293,878	104,580	61,271
1991	306,290	107,087	69,551
1992	312,226	108,218	73,871
1993	350,729	113,803	78,404

Source: Directorate General of Livestock Service (1994); Office of Statistics, Kabupaten (District) of Bandung (1995).

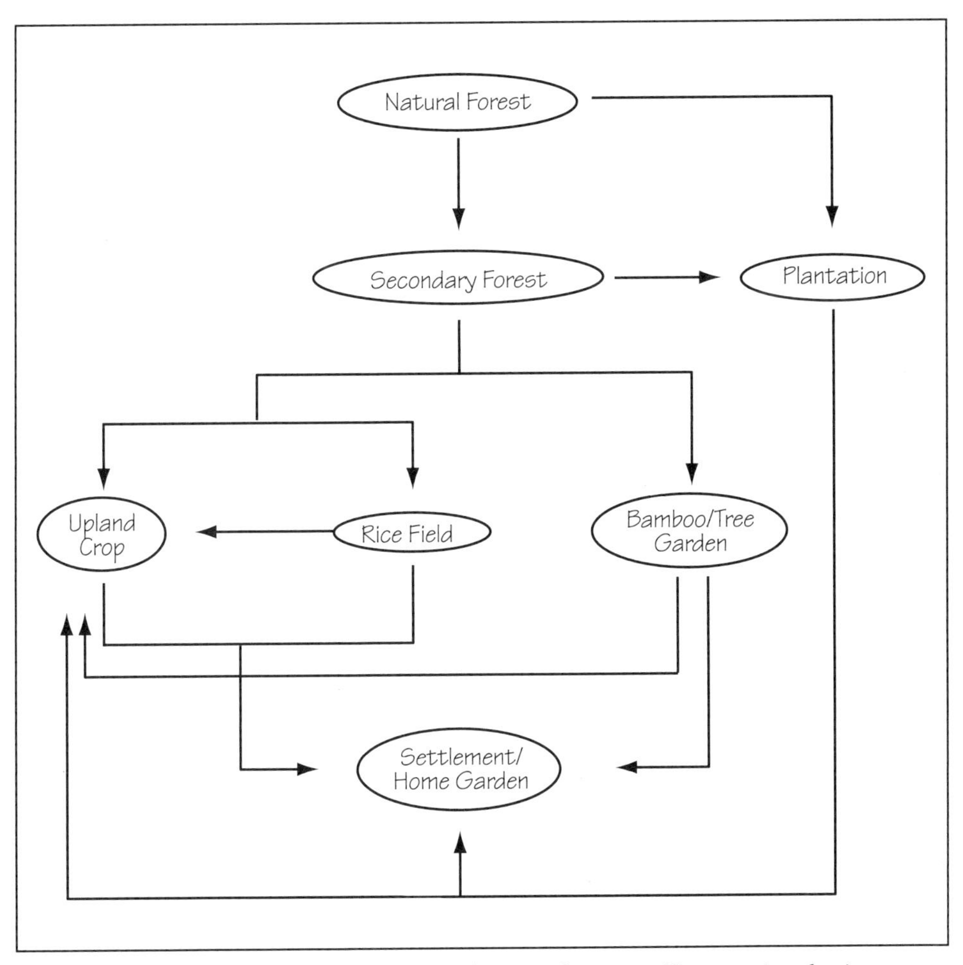

Figure 3.3 Historical changes in land use in the upper Citarum river basin (data from Soemarwoto and Soemarwoto, 1985)

late 1960s, have turned the watershed into a mosaic of natural and anthropogenic ecological communities with different compositions and structural vegetation patterns.

As figure 3.3 shows, human activity in the basin has resulted in the emergence of at least five major types of land use: tea and cinchona plantations, settlements/home gardens, upland cash crop gardens, rice fields, and bamboo/tree gardens.[2] The most significant changes have occurred since the 1970s, involving the rapid conversion to upland cash crop gardens of the bamboo/tree gardens, home gardens, rice fields, and private plantations. These land use changes have important implications for environmental conservation in the basin. Not only is habitat fragmentation occurring, but as bamboo/tree gardens and home gardens are converted to cash crop gardens, the complexity of the anthropogenic ecosystem is being dramatically reduced.

Dynamics of Biodiversity on the Agricultural Landscape

Studies were carried out in three villages in the upper Citarum river basin: Ranca Kasumba, Wangisagara, and Sukapura.[3] Ranca Kasumba is closest to an urban center, while Wangisagara lies in a zone of rapid agricultural and industrial transformation, and Sukapura lies on the periphery of economic and landscape changes. To compare the environmental conditions in the three study villages, the analysis focused on two types of land use: bamboo/tree gardens and home gardens.

BAMBOO/TREE GARDENS. The domesticated plants in bamboo/tree gardens can be placed in nine major categories based on their functions for local people, namely, vegetables, fruits, food, industrial material, medicinals, spices, ornamentals, fences, and construction materials and fuelwood. Overall, the number categorized as construction materials and fuelwood is the highest, followed by vegetables and fruits. Bamboo/tree gardens located far from the village tend to be planted with species used for construction materials and fuelwood, whereas those gardens located near the village are planted with fruit trees. This is because fruit trees need more intensive maintenance and closer monitoring to avoid theft of the fruits. A number of the species found in the bamboo/tree gardens can be utilized as either traditional medicines or vegetable; although medicinal plants are underutilized these days, people say, because of the distance from the village to the gardens and the availability of modern medicines.

Species inventories conducted in a sample of bamboo/tree gardens showed that the number of nondomesticated plant species tends to be higher than that of domesticated plants. The low intensity of management provides many opportunities for wild plants to grow. These inventories recorded a total of 237 species in the bamboo/tree gardens in the three study villages, classified into 194 genera and 64 families and consisting of 84 domesticated species and 153 nondomesticated species. The nondomesticated species are dominated by herbaceous plants, which generally are small, have good capacity for reproduction, and are well adapted to less favorable environmental conditions (Anderson 1976; Klingman 1982). Their dominance can be attributed largely to the characteristically low level of management in these gardens. Interviews with villagers indicate that none of the bamboo/tree garden owners in the three study villages did anything to control pests and only one or two persons in each of the villages fertilized or weeded their gardens (Parikesit and Yadikusumah 1997). Weeds are the most numerous category, represented by 101 species, and were followed by materials used for construction and fuelwood, at 63 species.

The diversity of species (domesticated as well as nondomesticated) in the bamboo/tree gardens is associated with a characteristic vertical structure, the complexity of which is often said to mimic the vertical structure of the natural forest. In many bamboo/tree gardens, in fact, a canopy gap between the upper and middle strata and/or between the middle and lowest strata can be distinctly observed. Often found in the upper strata are species that usually grow in natural forests such as *Ficus virens*, *Sterculia hantap*, and *Garcinia dulcis*. This complex stratification and associated high diversity of species are, in part, what makes bamboo/tree gardens a favorable habitat for some wild fauna, particularly birds (Erawan et al. 1997). A number of species of birds, some of them endangered and protected by conservation laws, use bamboo/tree gardens as both a food source and a nesting site. The palm civet *(Paradoxurus hermaphroditus)* is the most commonly found small mammal in bamboo/tree gardens. Its population in rural areas is otherwise diminishing due to the depletion of its habitat.

There is some intervillage variation in levels of diversity in the bamboo/tree gardens. A number of studies (Abdoellah 1990; Abdoellah et al. 1978; Karyono et al. 1978; Karyono, 1981; Widagda et al. 1984) suggest that rice cultivation and other types of intensive cultivation affect the plant diversity in associated land uses. The data on species composition in bamboo/tree gardens and home gardens in Ranca Kasumba, Sukapura, and Wangisagara support this hypothesis. Ranca Kasumba is in a paddy field area, and Sukapura is a vegetable-producing area. Intensive agricultural activities in these two villages are associated with lower mean numbers of domesticated species in bamboo/tree gardens than in the third village, Wangisagara. In the first two villages, the importance of bamboo/tree gardens as a source of income is overshadowed by paddy fields and cash crop gardens. The bamboo/tree gardens in these two villages mainly function to provide fuelwood, so the owners plant fewer species overall.

A comparison of plant diversity in the bamboo/tree gardens ranges from a high of 302 species in Wangisagara, to 296 species in Sukapura and 285 in Ranca Kasumba. In terms of domesticated species alone, Wangisagara also has the greatest species richness, with 196 species, while Sukapura and Ranca Kasumba have 159 and 171, respectively. Based on the average number of domesticated species per measured garden, Wangisagara again exhibits the highest numbers, with 8.4 in bamboo/tree gardens and 26.6 in home gardens. But the numbers for nondomesticated species show an opposite trend: Wangisagara has a total of just 106 wild species in its bamboo/tree gardens and home gardens, compared to 134 species in Sukapura and 109 in Ranca Kasumba.

HOME GARDENS. The total number of species found in our sample of home gardens in the three study villages exceeds that found in the bamboo/tree gardens: 269 species of plants belonging to 232 genera and 71 families. Among those species, 225 are domesticated and 44 are nondomesticated. The diversity of plants in the home gardens is actually greater than these figures suggest, as some species are represented by more than one variety.

The domesticated species found in the home gardens can be categorized in eight functional groups: ornamental, vegetable, fruit, food, industrial, medicinal, spice, and construction materials and fuelwood. The ornamental species are the most numerous (123), followed by vegetables (34) and fruits (23). The dominance of ornamental plants in the home gardens is due, in part, to the fact that they are situated next to the house and therefore reflect on the social status of the owner (Karyono et. al. 1978; Karyono 1981). Although ornamental plants dominate home gardens, some people also use them for nurseries, seed production, or market-oriented cash cropping as in some of the home gardens of Sukapura.

The majority of plants grown in the home gardens are small herbaceous plants. Their small size and shade tolerance (Tjitrosoepomo 1988) allows them to be planted in small pots and placed underneath the tree canopy or on the veranda. Multiple-purpose tree species used for shade, fruit, relish, and ornament are typically planted only in the larger home gardens, although sometimes one or two trees of a particularly useful species are also planted in small home gardens. Such trees are typically planted for their benefits to the household and not for market-oriented purposes.

The vertical structure of vegetation in the home gardens varies. Some gardens consist of only one stratum, but others show a complex vertical structure, ranging from herbs in the floor stratum to good-sized trees in the uppermost stratum. In some home gardens, trees may reach a height of 8 m or more. Variation in the vertical structure of home gardens is related to variation in garden function (Parikesit and Yadikusumah 1997). For instance, the simplest vertical structure—consisting of one stratum—is found in home gardens composed only of annual crops.

Variation in the complexity of the agroecosystem has strong implications for faunal diversity in the study area. In Wangisagara, where land use is more heterogeneous, floristic diversity is greater, the overall structural vegetation pattern is more complex, and the diversity of avifauna in bamboo/tree gardens and home gardens is higher than that in the other two study villages. Understanding the complex mosaic of land use systems and the role they play in ecosystem health and socioeconomic well-being is critical to a successful regional land management plan. Yet, as the following section indicates, the Indonesian government has so far failed to take such a regional view in its development planning. As a result, government policies aimed at forest management and agricultural development are in conflict and have failed to reach their goals.

Development Policies in the Upper Citarum River Basin

The Development and Planning Board of Bandung District has established a spatial planning strategy to manage economic growth for the period 1999–2019. Under this plan, the uppermost area of the Citarum river basin is targeted for development of smallholder agriculture, plantations, and conservation areas. These activities fall under the three major sectors being promoted by the Indonesian government in the upper basin: forest management, agricultural development, and dairy farming. In this section, these three land use systems are described and the role of state policy in managing each system is explained. In the subsequent section, the conflicts within macrolevel state policies are analyzed and the impact of these policies on regional conservation and landscape change are explored.

Forest Management

Forest areas located in the upper Citarum river basin are under the management of the Southern Bandung Forest Administration (Kesatuan Pemangkuan Hutan Bandung Selatan), a district-level institution of the State Forest Corporation. Of the total forest area managed by the Administration (55,423 ha), the majority is located in the upper basin.

Forests in the upper basin are divided into protection and production forests. The protection forest functions as a conservation area intended to maintain hydrological functions, while the production forest functions to produce income for the state. However, forest resources are also central to the economy of local households. The Indonesian government has tried to incorporate these local needs into forest management through two SFC programs, the *tumpang sari* (intercropping) program and the dairy cattle program. However, these programs, as described below, have failed to provide economic benefits to local people or to conserve natural resources.

Prior to the mid-1980s, the SFC involved local people in managing the state's production forests by giving them access to arable forestland under the *tumpang sari* program. Local farmers were allowed to plant agricultural crops between rows of tree species for one to three years. Farmers were obliged to take care of the SFC's newly planted trees as the quid pro quo for temporary use rights to cultivate the land. The program was intended to alleviate what the state saw as illegal cultivation of land within the production forests by actively involving local people in forest management and by providing them with a temporary source of agricultural land. This program was terminated in 1986,[4] however, when local officials feared that it was actually resulting in increased forest encroachment and more flooding and downstream erosion (see Hardjono 1990; and Sulaeman 1997).

The termination of the *tumpang sari* program did not stop forest encroachment. The high rate of local landlessness meant that people continued to look to the forests for new land for agriculture. Additionally, flooding in downstream areas was not necessarily linked to the *tumpang sari* program at all but was more likely the result of both the expansion of agriculture and the growth of industrial activity in the lower reaches of the basin. Thus, terminating the *tumpang sari* program did not alleviate the problem of forest encroachment and flooding because it did not create the problem in the first place. Rather, the broader regional shortage of agricultural land and rapid industrial development were responsible for forest conversion and downstream flooding. Regional officials did not consider these issues of regional development and underdevelopment in either the formation or the termination of the *tumpang sari* program.

Agricultural Development

Agriculture development in Indonesia has been given priority in the national development programs since independence from the Dutch. The central government has consistently placed heavy emphasis on development of the agricultural sector through extensification and intensification programs, such as those associated with the Green Revolution. The government claimed success for this policy when in 1984 the country was able to rely solely on internal rice production for national consumption, without importing any rice from other countries.

In the upper Citarum river basin, the Bandung District Agricultural Office (Dinas Pertanian Kabupaten Bandung) determined that the uppermost part of basin should be a market-oriented production center for vegetables, whereas the lower area should be a production center for rice. The success of the government's efforts to develop cash crop cultivation on the upper Citarum is demonstrated in the annual production statistics, which show a dramatic increase in vegetable crops (such as red onions, potatoes, carrots, and cabbage) during the period 1990–96. The office reported that total Bandung District production of red onions, which was just 27,111 tons in 1990, had increased to 30,788 tons in 1995 and 57,680 tons in 1996. District-wide production of potatoes, which was only 18,900 tons in 1990, rose to a remarkable 226,208 tons by 1996.

Despite the national success story in rice self-sufficiency and the regional success story in rapid development of cash crops, unexpected and negative effects of this agricultural development have been widely reported. In fact, the achievements of the Green Revolution have exacted environmental costs such as the loss of local cultivars due to monovarietal cropping (see Frossard, this volume), decreased soil fertility, recurrent pest outbreaks (Winarto 2004), and soil erosion (Griffon 1997). The Green Revolution also has spawned a host of social problems. Changes in the agricultural labor structure and attendant rural impoverishment and increasing urbanization are among the major problems that have accompanied the intensive development of the agricultural system. It is also clear that the ecological and social impacts of this agricultural development have yet to be fully explored.

Dairy Farming

Dairy farming has received considerable developmental support in Indonesia for two reasons. First, it is supposed to contribute to the national production of milk and thereby conserve foreign exchange; and, second, it is supposed to alleviate poverty by providing alternative sources of income for rural people. In fact, raising dairy cattle is a lucrative business. With only 2 or 3 productive cattle, a farmer can earn enough income to support a family of five persons above the poverty line.[5] As a result, the number of cattle raised by farmers has increased significantly in the last two decades. Data from the Southern Bandung Dairy Farmers Cooperative (Koperasi Peternak Sapi Perah Bandung Selatan, or KPBS) show that the number of dairy cattle in southern Bandung increased from just 2,608 in 1969 to 21,000 by 1994. The KPBS planned to further increase the cattle population. In the 1980s, a study estimated that the KPBS region could support from 35,000 to 45,000 cattle,[6] officials at the Dinas Peternakan Kabupaten Bandung (Bandung District Husbandry Office) are using this estimate as a target for future development of the dairy cattle industry in southern Bandung.

Development of the dairy industry in southern Bandung has been hampered by several factors. The most significant constraint is the insufficient supply of grass, particularly in the dry season. Dairy farmers have to search and compete with other farmers for grass in the forest, and some must search for grass as far away as Bandung municipality, 40 km distant (see Gunawan, Abdoellah, and Parnawan 1997).

In 1986, the KPBS, in cooperation with the State Forest Corporation began to permit dairy farmers to plant fodder grasses in the state forests. The SFC adopted this fodder program in the wake of the perceived failure of the *tumpang sari* program. The KPBS coordinated the grass-planting activities of its members, but it refused to take responsibility when some of these members converted the land they had planted with grass to the cultivation of cash crops. The KPBS argued that it was the responsibility of the farmers to follow the program's goal of planting only grass for cattle fodder. As a result of this conflict, the SFC terminated this program as well, although it tacitly continued to tolerate local dairy farmers who planted only grass in the forest area, as this appeared to have no detrimental effects on the forest environment.

The difficulties in getting grass, particularly during the dry season, have forced some erstwhile dairy farmers to sell their cattle and invest their money in other economic activities. This has been the primary factor responsible for the decline in the dairy cattle population in recent years. Additionally, profits from the industry shrank when the price of cattle feed increased while milk prices remained relatively stagnant. In light of the smaller profit margin, officials of the Bandung District Husbandry Office believe that the future of the dairy industry is in doubt, especially because the market price of Indonesian farmers' milk has generally exceeded that of imported milk. When the Fair Trade Agreement of the Association of Southeast Asian Nations (ASEAN) took effect in 2003, there was no longer any protection for local dairy farmers and the milk-processing industry is now allowed to freely procure milk from any source at competitive prices.[7]

Despite these negative developments for the dairy industry, the recent monetary crisis has given dairy farmers at least a temporary respite. The depreciation of the Indonesian rupiah raised the price of imported milk above that of domestic milk even after substantial increases in local milk prices. It is thought that in the foreseeable future local milk prices will continue to be lower than those of imported milk (*Kompas,* February 21, 1998). This condition is beneficial to local dairy farmers, and this may lead to temporary increases in the number of dairy cattle.

Conflicts and Consequences of Government Policy

As demonstrated in the preceding pages, the changes in the ecology of the upper Citarum river basin have been caused not only by changes in local patterns of land use but also by the uninformed and uncoordinated formation and implementation of government policies. The lack of cohesion among government agencies overseeing forest management, agricultural development, and dairy farming is responsible for much of the vulnerability of the ecosystem of the upper basin. Two policies in particular warrant close examination in this regard. The first is the termination of the *tumpang sari* program, which occurred at the same time as the takeoff in the development of the dairy industry. The second is the governmental emphasis on converting home gardens and bamboo/tree gardens to strictly market-oriented agriculture rather than conserving the complex traditional agricultural systems that better promote biodiversity. In the following sections, these two policies are examined in detail.

The "Tumpang Sari" Program and the Dairy Industry

A critical decision made by the government concerning forest management in the upper Citarum river basin was the termination of the *tumpang sari* program in 1986. The program was canceled in the belief that this would protect the forest from local encroachment and restore its hydrological functions. The termination of the program was in conflict with other governmental policies, specifically the plan of the Dinas Peternakan (Animal Husbandry Administration) and the KPBS to develop the dairy cattle industry, which was also dependent on the use of forest resources. Studies (e.g., Gunawan et al. 1997) show that development of the dairy industry in Taruma Jaya, for example (see fig. 3.1) is actually subsidized by the use of resources in the Mount Wayang forest. With high rates of landlessness and a shortage of fodder, the dairy industry is dependent on forage in the forests. It would have been very difficult to develop the dairy industry if farmers were not given access to the forest for forage. Local policymakers failed to see the conflict in terminating the *tumpang sari* program, in the belief that it was responsible for forest encroachment, and simultaneously supporting the growth of a dairy industry, which was also dependent on forest resources.

Even more important, governmental support for developing a larger dairy industry was given without clear knowledge of the carrying capacity of the area for cattle. With a lack of coordination between the SFC and the KPBS, the enlarged dairy industry has in fact heightened pressure on forest resources (Gunawan et al. 1997). Not only do dairy farmers rely on the forest for forage for their cattle, but they regularly take timber from the forest to build stables. Although the government promoted the development of the dairy industry as a remedy for the problems associated with the *tumpang sari* program, this development has actually had the opposite effect.

The termination of the *tumpang sari* program and support for the development of dairy farming have not succeeded in protecting the forests from encroachment. In fact, encroachment on to the forest in the upper Citarum has increased at a precipitous rate. By June 1997, it was reported that about 1,500 people were cultivating land in the state forest surrounding Mount Wayang. Until recently, this information on forest encroachment, while known to local-level officials, was not even reported to the higher, provincial officials of the SFC.

Agricultural Development versus Bamboo/Tree Garden Conservation

Another significant conflict exists between the policies of the government's Agricultural Office and the Land Conservation and Forestry Office. The Agricultural Office believes that increased agricultural production depends on replacing subsistence-oriented land uses with more market-oriented ones. As a result, it is promoting conversion of bamboo/tree gardens into cash crop gardens. This program of conversion is in direct conflict with the Land Conservation and Forestry Office's announced policy of bamboo/tree garden conservation. This policy is based on Peraturan Daerah (Regional Regulation) no. 11/1995, which regulated all cutting of bamboo. Implementation of the policy and the supporting law has been nonexistent, however. During 1997–98 the Land Conservation and Forestry Office was unable to carry out any bamboo/tree garden conservation projects in the upper Citarum river basin. Its officials argue that to develop this kind of agroforestry system in an area dominated by cash crop agriculture is very difficult. Their support for the conservation of bamboo/tree gardens is undermined both by the reality of the high profits that can be obtained from cash crop cultivation and by the Agricultural Office's explicit efforts to promote such cultivation.

The rapid conversion of bamboo/tree gardens to cash crop gardens has several implications for the ecological stability of the region. First, the new market-oriented crops and high-yielding varieties that have been introduced into the upper basin have displaced many traditional varieties and therefore have contributed to the general biological impoverishment of the wider ecosystem. The loss of genetic diversity has produced a drastic change in the population structure of crops, and it has both heightened vulnerability to pests and created environmental stresses through the excessive use of pesticides.

Second, there has been a loss not only of traditional crop varieties and species but of traditional land use systems. Human activity in the upper basin historically augmented landscape diversity, as forests were converted to at least five different major types of use (namely, tea and cinchona plantations, home gardens, cash crop gardens, bamboo/tree gardens, and rice fields). In recent years, however, landscape diversity has been on the decline and homogenization of the landscape has been taking place. As a result of the growing emphasis on cash crop cultivation, bamboo/tree gardens (which have less economic value to farmers) have dramatically declined in extent and have even disappeared from some of the villages in the study area.

Third, the conversion of bamboo/tree gardens to cash crop gardens has had consequences for the ecological structure and function of the wider landscape. The loss of the stratified canopy structure and species diversity of the bamboo/tree gardens means a loss of the gardens' multiple functions in moderating microclimates and conserving soil and water. By way of comparison, cash crop gardens, with their low species diversity and lack of litter on the ground, are much less successful in conserving soil, as can be seen in some parts of the study region, especially on steep slopes.

Key Issues Behind Forest Encroachment

The Indonesian government's efforts to alleviate pressure on forest resources and manage lands sustainably in the upper Citarum river basin have failed in part because they have ignored four key socioeconomic issues: (1) high population density, high rates of landlessness, and a regionwide dependency on agriculture; (2) local dependence on fuelwood and timber for construction; (3) external market demand for forest resources; and (4) weak enforcement of existing forestry laws. In this section each of these issues are briefly explored.

Population density is a prominent factor in the transformation of the landscape and the conditions of biodiversity in the upper Citarum river basin. The overall population density of the upper basin is 519 persons per sq km, which is actually lower than the average population density for upland areas of Java, which is 671 persons per sq km. However, in reality the Citarum density is much higher than this. If state forests and plantations are excluded from the calculation of land available for agriculture, the population density of the villages in the upper basin ranges between 2,000 and 10,000 persons per sq km. Population density in the territory of Taruma Jaya in the foothills of Mount Wayang, for example, is nearly 10,000 persons per sq km.

The impacts of high population density are exacerbated by inequalities in land distribution. The percentage of landowning households in the three study villages in the upper basin varies between just 10 and 40 percent,[8] with the lowest percentage of landowning households found nearest Mount Wayang. Furthermore, the majority of the landowners have only one plot of land, averaging less than 0.25 ha in area. Even though the majority of the population is landless, people are still dependent on the agricultural sector for their livelihood. Consequently, they have no alternative other than encroaching on the forest and cultivating forestland.

Not only is forestland needed for agriculture, but fuelwood is in high demand in this region. Survey data indicate that 82.5 percent of the villagers still use fuelwood to meet their daily energy needs. The villages above 1,200 m in altitude have a relatively cold climate and fuelwood is their main source of energy. In Sukapura, for example, villagers rely on fuelwood not only for cooking but for heating and such tasks as drying grain. Given the multiple uses of fuelwood, this kind of energy source is not easily replaceable with alternative energy sources such as kerosene or natural gas, even if these were affordable. Some villagers have planted fast-growing tree species to meet their energy needs, but most rely on naturally growing trees. The majority of the people in the foothills of Mount Wayang rely on timber from its forests for fuelwood. The environmental impact of fuelwood collection depends on its intensity. Up to a certain point, collection of fuelwood may have little impact. At high harvest rates, however, removal of even deadwood implies a reduction of food and substratum for large groups of species (McNeely et al. 1995). In addition to fuelwood, the villagers are also dependent on the forest for timber for house construction. The full impact of the collection of timber for fuel and construction purposes has yet to be studied.

State support for the dairy industry has unintentionally contributed to the problem of forest degradation by increasing demand on the forest for fuelwood, used to prepare cattle feed as well as process milk, and timber used in the construction of cattle pens and stables. Survey data indicate that at least half of all dairy farmers take wood for the construction of stables. Not only do dairy farmers take wood for these purposes, but whenever grass is difficult to get, especially in the dry season, many lop tree branches to feed their cattle. Thus, as the dairy industry grows it places even more pressure on the forests.

Aside from the development of agriculture and animal husbandry, there are other sources of pressure on the flora and fauna of the upper Citarum river basin. For example, this study found that the prickly ash *(Zanthoxylum* sp.) is now becoming very rare in the Mount Wayang forest because of intensive harvesting by both local people and outsiders for the ornamental bonsai market (see Parikesit et al. 1997). Both locals and outsiders also hunt birds with distinctive plumage and/or voices for sale in urban markets. Birds such as the black-naped oriole *(Oriolus chinensis)*; the orange-fronted barbet *(Megalaima armillaris*, which is protected and endemic to Java and Bali); the lesser shortwing *(Brachypterix leucophaeus)*; the white-browed shortwing (*Brachypterix montana*); and the Sunda blue robin *(Ciclidium diana)* are so sought after that they are now rarely found in the Mount Wayang forest (see Erawan et al. 1997).[9]

The overexploitation of natural resources is made possible in part by inadequate enforcement of the law. To begin with, there are insufficient forest guards available to police the forested areas. There are only six guards assigned to control the forests surrounding Mount Wayang, Mount Windu, Mount Bedil, and Mount Malabar, a territory that measures over 2,700 ha (see fig. 3.1). The guards argue that it is difficult for them to adequately police such a vast forest area. But even when illegal loggers, for example, are apprehended by the forest guards, enforcement of conservation laws may be undermined by corruption. It is often reported that those apprehended for illegal logging are released after bribing the officials in charge before their cases are brought to court. In another example of political corruption, local people reported that prior to the 1997 elections they were promised that they could cultivate forestland if they voted for a certain political party. Such corruption and weak law enforcement have greatly exacerbated the problem of forest encroachment and degradation.

Recommendations for the Future Management of the Upper Citarum River Basin

In this chapter, we have provided a broad overview of the complex linkages between environmental conservation, local socioeconomic conditions, and the failure of state policy initiatives. It emphasizes that the linkages between policy centers and the peripheries, and between broad state agendas and local village needs, need to be more clearly articulated in order to better design state policies for both local people and the natural resources on which they depend. Macrolevel policy needs to incorporate a much fuller understanding of microlevel ecological and socioeconomic conditions.

Significant institutional reorientation must occur if we are to achieve sustainable management of the natural resources of the Citarum river basin. The first step in this direction is to educate government officials concerning concepts such as sustainable development and resource use, the value of biodiversity conservation, and the importance of complex and heterogeneous land management systems such as those traditionally found in the region. The principles of biodiversity conservation are not clearly understood by the officials in the region, and many claim that they have never factored them into their decision-making processes. Local government officials and policymakers also need a better understanding of landscape-level dynamics. The upper Citarum river basin comprises a number of integrated land uses, which cannot be managed separately. A lack of integration between the polices of different government offices is responsible for some of the current problems in the basin. If the basin were seen as a single management unit, the relationships among forest encroachment, development of the cattle industry, landlessness, and the excessive dependence on agriculture would be easier to comprehend. Integrated planning will require more dialogue and better information flows among the diverse government agencies concerned with resource management in the region as well as clear communication between local officials and central policymakers.

The last three decades of development in the upper Citarum river basin show that whereas cash crop cultivation and dairy farming appear economically beneficial these developments are not sustainable at current rates. Therefore, a second major conceptual reorientation is needed to emphasize long-term sustainability over short-term productivity. For example, bamboo/tree gardens play an important role in sustaining biodiversity and landscape heterogeneity in the Citarum watershed (Abdoellah et al. 1997; Parikesit and Yadikusumah 1997; Erawan et al. 1997), but the immediate economic benefits of gardens are low, particularly compared to more intensive agricultural systems. In order to prevent bamboo/tree gardens from disappearing and in an effort to slow landscape homogenization, it will be necessary to improve the direct economic value of these gardens. One possibility would be the development of bamboo/tree gardens as sites for tourism, village home stays (which are increasingly popular in Southeast Asia), and/or development of bamboo-related handicrafts. Each of these activities could potentially generate substantial income for farmers while preserving both the sustainability of the bamboo/tree gardens and their valuable contribution to landscape heterogeneity.

In order to facilitate this reorientation away from quick economic returns and toward long-term sustainability, the government should also provide appropriate disincentives. For example, increased taxes could be levied on land converted from traditional uses to cash crop gardens, and tax incentives could be given to farmers who conserve existing bamboo/tree gardens or plant new ones. Equally important, in an effort to emphasize long-term environmental sustainability over short-term economic gains farmers need to be better educated about the negative impact of heavy chemical inputs. Farmers, dependent on agricultural profits, will not reduce the use of chemical inputs without economic support from the government. The development of markets for pesticide-free agricultural products will help, particularly if this is done in conjunction with an extension program promoting less reliance on chemical pesticides and more reliance on integrated pest management systems (see Winarto 2004). Concerns surrounding pesticide use are successfully driving the development of organic vegetable markets in Europe and the United States, and there is a potential to develop similar markets in Indonesia.

In order to reorient development in the upper Citarum river basin toward long-term economic benefits and sustainable environment conditions over short-term profits, policymakers and governmental institutions must recognize that the search for increased economic benefits and agricultural self-sufficiency has threatened the long-term sustainability of the environment and in consequence the economic viability of the region as well. What is needed is an ecosystem approach to understanding the management of natural resources within the context of local economies. Such an effort needs to function at multiple scales, so that the microlevel details of local socioeconomics and ecological biodiversity are placed within the regional scale of landscape change and economic development. This kind of ecosystem-level approach will enable us, for the first time, to simultaneously analyze the challenges of resource management at the farm, village, river basin, and regional levels. This should give us greater insight than we have ever had into the question of why management policies issued by the center for the periphery fail and how they can be revised so as to succeed.

Notes

1 Soesilo et al. (1998) report that in the preceding decade 688 ha of forest in the upper basin were cleared, 28,684 ha of bamboo/tree gardens and home gardens were converted to settlement and industrial uses, and industrial land use increased by 38,620 ha.

2 People from different villages often use different terms for the same land use type. For example, the terms *kebon awi*, *kebon tatangkalan*, and *bojong* are used in different villages to describe a piece of land planted with bamboo and fruit trees—hereafter called bamboo/tree gardens.

3 This study uses an ecosystem approach in assessing biodiversity conservation. Barbault and Sastrapradja (1995) point out that to face the challenge of the loss and management of biodiversity, a spatial- and ecosystem-oriented approach appears to be more useful than an accumulation of species-centered studies.

4 The SFC initiated the move to terminate the *tumpang sari* program in the upper basin in a letter to the provincial government of West Java (Letter no. O22.7/Dir, July 28, 1986). Based on this initiative the governor of West Java issued an order to terminate the program (Letter no. 521.22/8066/Binprod, September 16, 1986). After the program was terminated, the SFC shifted to hiring daily laborers to plant and care for its trees.

5 This was less true after July 1997, when Indonesia's economy crashed due to the regional monetary crisis.

6 Interview with a top official of the KPBS, reported in *Pikiran Rakyat*, October 17, 1996.

7 Prior to this, Indonesia's milk-processing industry had been obliged to purchase a set amount of milk from local farmers before it could import any, a measure intended to protect the country's dairy farmers from foreign competition.

8 This is in accordance with the wider picture of landownership in West Java, which shows that more than 50 percent of the people are landless.

9 The introduction of air rifles into the area in the late 1970s is considered to be another significant factor in the diminishment of avian diversity in the upper basin.

4

In Field or Freezer? Some Thoughts on Genetic Diversity Maintenance in Rice

David Frossard

There is a cynical (if nonetheless valid) journalist's proverb that says, "Freedom of the press belongs only to those who own the presses." By the same token, the freedom to breed new strains of rice—say, to combat the rise of new pests or adapt to climatic shifts—rests in the hands of those who control the old rice varieties, the raw parent material for new generations of crops. In Asia, that freedom is enjoyed mostly by national, international, and commercial crop research centers that hold in cold storage thousands of varieties of rice seed no longer found in farmers' fields.

While the ongoing removal of rice farmers from the process of varietal development (and their conversion to passive recipients of technology and seeds produced by scientists) has had profound cultural and economic effects throughout Asia, the effects of this new genetic regime on rice biodiversity have been no less profound. Tens of thousands of species previously grown in farmers' fields became a few thousand, then a few hundred. The very idea of *farmer* maintenance of rice genetic diversity has been dismissed—at least by many professional participants in the crop-development industry—as unworkable, improbable, and ultimately virtually *unthinkable*.

However, many farmers—including members of a group in the Philippines known as MASIPAG, described below in a brief case study—are unwilling to give up their age-old participation in the production of new and better varieties of rice (see n. 14). More to the point, they are unwilling simply to accept scientific rice-growing technologies that are unprofitable and disadvantageous for themselves and their communities.

The dichotomy between cold storage of rice genetic material in the laboratory and living conservation of genes in the field reflects not simply an esthetic difficulty—nor even a scientific one—but involves larger questions of farmer rights, ownership of genetic materials, colonialism, racism, and finally *scientism* (the unwarranted privileging of one particular form of scientific discourse over other ways of knowing). The struggle to control rice germplasm in Asia is thus indicative of a larger struggle by farmers to find a (social, political, economic, and cultural) place for themselves in a rapidly changing world order.

Into the Freezer: "Preserving" Rice Genetic Diversity

As we begin a new millennium, humans are slowly becoming aware that our own survival as a species may depend on the continuing genetic diversity of millions of plant and animal species that form a complex and incompletely understood planetary ecosystem. Yet rain forests, for example, continue to be destroyed at a rate of 1.8 percent per year, and at least 0.2 to 0.3 percent of species living in those forest tracts—at least four thousand unique plant and animal species—each year perish forever (Ehrlich and Wilson 1993, xiii). Worse, while these estimates are chilling, they are nonetheless *conservative*. The true rate of extinction may in fact be "two or three orders of magnitude greater" (xiii). Many thousands of undiscovered plants and animals with potential medicinal or food value (or crucial ecological properties we have not yet begun to understand) have disappeared; many more will follow.

Despite this record of destruction, humans have paid at least passing attention to the relatively few species that most directly make our lives on the planet possible. Several dozen major plant species, including potatoes, corn, wheat, and rice, the latter being arguably the world's single most important food, have been selectively "improved" by humans for millennia. Through meticulous selection of promising varieties, farmers slowly transformed the relatively undistinguished wild progenitors of these crops into genetically diverse food sources, productive in myriad local soil types and microclimates, and resistant to specific local pests.

A generation ago, tens of thousands of rice varieties were grown by farmers in Asia and Africa.[1] Today scientists at the Philippines-based International Rice Research Institute (IRRI), the world's premiere rice-research center hold seeds of about 75,000 different varieties of Asian rice, *Oryza sativa*, in long-term cold storage for use in scientific crossbreeding programs. The institute and other germplasm storage centers spend millions of dollars annually to preserve millennia of human ingenuity in the form of indigenous rice seed stocks (IRRI 1993).[2]

At the same time, due largely to the massive international scientific effort to "improve" this crucial plant species, biodiversity of rice in the *paddy field* (as opposed to the *freezer*) has *decreased* precipitously in a single generation. Consistent with a scientific crop-breeding program that transforms the social location of seed germplasm (and political control over it), work done at IRRI and similar scientific institutions was the single major cause of that in situ diversity decrease.[3]

The political and scientific processes by means of which genes flowed away from the field and into the freezer are well documented in the social science literature and need not be revisited here at length.[4] However, one might note the appearance of the first so-called Green Revolution variety, a high-yield, input-intensive rice which was known as IR8, released by IRRI in 1966. It was so productive, and at first so profitable, that farmers flocked to this new "miracle" rice, leaving many indigenous varieties unplanted.[5]

A later IRRI release known as IR36 was the most widely grown rice in history, covering 11 million hectares in Asia in its heyday and about 60 percent of all rice production in the Philippines, Indonesia, and Vietnam at that time (Salazar 1992; see also Jackson 1995; and IRRI 1992). But, while IR36 gave excellent yields and resisted pests well for several years, its very "success" meant that many local landraces were driven to extinction in the field.

Today, after almost forty years of Green Revolution rice growing in Asia, thousands of varieties once grown by farmers are found only in cold storage in places such as IRRI, while a few score high-yield, chemical-intensive, Green Revolution varieties make up the vast bulk of all rice grown. The indigenous rice developed by farmers has become, in many cases, simply the raw material for future *scientific* rice production. The farmers who were once the wellspring of genetic diversity in rice are now, by and large, merely the recipients of scientific largesse, buying new varieties regularly on the open market rather than producing new strains themselves.[6]

A sad state of affairs, perhaps, but is this really a problem? Is criticism of this arrangement simply nostalgic longing for a style of rice growing, and a style of life, now all but gone? To be fair, centralized, scientific control of rice germplasm has some arguably positive aspects.

For instance, the IRRI germplasm archive (duplicated at the National Seed Storage Laboratory in Fort Collins, Colorado, and to some degree in various national collections) provides insurance against disastrous genetic loss. In one of its proudest accomplishments, IRRI successfully reintroduced rice strains that had disappeared in the chaos of Cambodia's killing fields in the 1970s.[7] And from the point of view of plant breeders, centralization of genetic material allows scientists efficient access to a wide variety of potential parent stock for manufacturing new strains (e.g., IR36, for example, contained genes from fifteen indigenous rice varieties and one "wild" rice relative, *O. nivara).*

Scientific manipulation of rice genes in places such as IRRI may yet give rise to wildly better new kinds of rice. Super-high-yielding strains based on an entirely new "architecture" of the rice plant, perennial rice that needs no replanting, and a self-fertilizing rice that can "fix" its own nitrogen from the air are possibilities periodically mentioned in the IRRI literature (although the environmental side effects of such radically new plants are uncertain and by no means guaranteed to be positive or predictable). Finally, control of seeds by nonprofit international centers such as IRRI may serve as a counterweight to multinational seed corporations seeking to privatize and patent seed stocks.

Certainly, the official IRRI policy on intellectual property rights and rice genetic resources (quoted in Jackson 1995, 273) is hopeful. It states that: (1) The rice genetic resources maintained in the gene bank at IRRI are held in trust for the world community; (2) IRRI adheres to the principle of unrestricted availability of the rice genetic resources it holds in trust including related information; (3) IRRI will not protect the rice genetic resources it holds in trust by means of any form of intellectual property protection; (4) IRRI is opposed to the application of patent legislation to plant genetic resources (genotypes and/or genes) held in trust; and (5) the rice genetic resources held in trust by IRRI will be made available on the understanding that the recipients will take no steps that restrict their further availability to other interested parties. Nonetheless, the trend is probably in the direction of private control of such resources. For example, Shiva (1994) lists twenty-one multinational companies actively collecting and screening plants and other natural products for commercial gain.

In any case, while Green Revolution agriculture has enabled the planet to feed most of its citizens for a generation, even with a steadily increasing world population, there are signs that the limits of this technology are near. The land frontier has been reached in many parts of Asia, and paddy area has remained static or has decreased since 1980 (IRRI 1992). Not only is more prime land unlikely to be devoted to rice, but many prime rice-growing areas disappear each year under the pressure of urbanization. Any increase in hectarage must come in marginal lands (where specific, *locally adapted* rice varieties are likely to be most productive).

Elaborate irrigation systems, which account for perhaps one-quarter of the increased yields attributed to the Green Revolution (Herdt and Capule 1983, table 10), are no longer a top priority in national budgets. New systems will remain unbuilt; old ones will deteriorate (Khush 1995). Pesticides and herbicides pose an increased threat to human and livestock health, even as pests build new resistance to these chemicals (see, e.g., Loevinsohn 1987). Biodiversity in the field—as an alternative to increased pesticide use—will become even more crucial in the future.

Ominously, in many rice-growing areas (especially those that have been continuously flooded for decades under a multiple-cropping regime) the rate of increase in rice yields has declined; in others, *absolute* yields are down (Cassman et al. 1995). More and more nitrogen fertilizer is used at great expense to maintain yields or minimize losses. The annual rate of growth in rice production has dropped from 2.8 percent in 1975–85 to 1.2 percent in 1985–92—*below the rate of population increase* (Khush 1995). If present trends hold, a rice deficit will soon exist worldwide. And, according to one controversial estimate, China alone will soon consume all of the world's surplus stores of rice (Brown 1996).

Despite the doubling of rice yields with chemical-intensive Green Revolution technologies in the past generation, "the magnitude of the increase in food output required in the next 30 years will be larger than the increase of the past 30 years. Whether farmers can continue to sustain yield increases faster than the rise in demand from population growth remains the key issue," according to Cassman and Pingali (1995a, 299).[8]

In sum, warns IRRI principal plant breeder Gurdev Khush, "the increased demand for rice ... will have to be met from less land, with less water, less labor and less pesticides" (1995, 281). Some researchers already look to a "post-Green Revolution era" in the near future (Pingali, Moya, and Velasco 1990; Cassman and Pingali 1995a, 305).

The institutional response to this serious situation has been clear for some time. Scientists at IRRI and elsewhere in the international agricultural research network will attempt to increase yield ceilings by 30 percent, 50 percent, and more by "reengineering" the rice plant. A new "ultra-high-yielding" rice type currently being prepared at IRRI will have a modified architecture: fewer tillers (stalks) than today's plants, all of them bearing grain; very quick maturation; heavier grains with better milling quality; better germination rates; and more vigorous seedlings (IRRI 1991a).

Biotechnology will figure strongly in the production of new kinds of "superrice." The Rockefeller Foundation alone spent $33 million from 1986 to 1990 to support biotechnology research. One-eighth of the amount designated for work on rice went to IRRI (IRRI 1991b). Soon genes from species other than rice (even nonplant species) may make their way into rice through genetic engineering.

Yet, despite the huge sums and cutting-edge science now applied to the search for next generation rices, it is difficult to see how a few varieties of superrice alone can solve the problems enumerated above. Indeed, unless the new varieties are unusually disease resistant, the risk of planting these few monogenetic strains on millions of hectares is chilling. While a new superrice may raise yields, it is also likely to lower genetic diversity in the field—perhaps to critical and dangerous levels. Ironically, the more "successful" a single type of rice is—the better its distribution among farmers and nations and the greater its popularity—the more potential exists for disaster. Further, almost 40 percent of the area planted to rice is in particular rainfed lowland, upland, or flood-prone areas—unfavorable environments in which "modern" varieties have had very little success.

No one variety of superrice can suffice for the very different conditions represented here. It is likely that in the future increased yields will depend on knowledge-intensive farming and myriad rice varieties tailored to local microenvironments. Production of these many varieties is a job for which a centralized organization such as IRRI is institutionally unsuited. And yet the possibility that local farmers, working on their own or with the help of sympathetic outsiders, could be part of the solution is rarely broached. The question should be asked: Why have farmers—the original developers of rice—become virtually invisible in this process today? And how does in situ maintenance of rice genetic diversity promise to both bring farmers back into the equation *and* ameliorate some of the difficulties listed above?

How Are You Gonna Keep (Seeds) Down on the Farm?

In situ conservation has important features not shared with laboratory storage of genes. For instance, while farmers' fields produce *new* genetic mixtures every season with cross-fertilization between nearby varieties, institutional cold storage is merely a "snapshot" of rice diversity—static, unchanging, and immune to the pressures of natural selection. Without the influx of new varieties from farmers' fields, diversity expansion in the laboratory comes only from a relatively small number of artificial crosses made by scientists (small in comparison with the inestimable number of crosses that would have taken place naturally in farmers' fields).

Similarly, farmer experimentation invariably produces rice strains appropriate to particular local soil and climate conditions—by definition, rice that doesn't measure up is not grown the following season or (when farmer seed exchange is practiced) perhaps it is grown successfully somewhere else under different conditions. In comparison, an organization such as IRRI is limited in its ability to tailor rice varieties to a wide variety of growing conditions, settling instead for broadly adaptable varieties not necessarily perfectly suited to a particular place. Even IRRI's partnerships with national seed production centers offer only a relatively small improvement in the diversity equation and provide farmers with only a relatively few more locally adapted varieties.

Genetic diversity itself is a hedge against pest problems and is a key way for farmers to minimize the risk of crop failure—essentially the lesson taught by the Irish Potato Famine (Fowler and Mooney 1990, ch. 3). When rice farmers grow hundreds of different varieties in a limited area (as they continue to do in some remote upland areas where new rice types have made few inroads), pest infestations are generally restricted to small areas of genetically vulnerable rice. In this situation, most of the (highly diverse) crop survives and pests tend to be self-limiting. However, in the case of extensive monogenetic monocrops, pests may successfully attack large areas, reducing farmers to a choice between large-scale spraying of pesticides or disaster.[9] Of course, scientifically developed seeds are now often highly pest resistant. Over time, however, pests evolve (or plant genes drift) and new, more-resistant rice types must be introduced, beginning the cycle of risk again.[10]

Finally, production of new rice varieties by farmers themselves may have empowering political effects. Depending on the level of "democratic space" available to them in their particular situation, farmers may be able to successfully assert ownership rights to the genes present in their fields. Such potential is suggested by the *International Code of Conduct for Plant Germplasm Collecting and Transfer* (FAO 1994) recently advanced by the Food and Agriculture Organization of the United Nations. The voluntary code, which enumerates the rights and responsibilities of collectors and owners of genetic resources, suggests that institutional users of germplasm should "consider providing some form of compensation for the benefits derived from the use of germplasm," including providing farmer access to "new, improved varieties and other products, on mutually agreed terms" and "any other appropriate support for farmers and communities for conservation of indigenous germplasm" (18).

As equivocal as this document may be, it nevertheless provides a framework within which farmers can claim recompense for the products of their experimentation. In contrast, seed held in cold storage is unlikely to provide farmers with as compelling a claim (if indeed the original farmer-owners of seeds collected decades ago might still be identified—an impossibility in many cases).

Farmers who produce their own varieties may assert a stronger claim to profits deriving directly from the sale of farmer-derived seeds in the wider marketplace. Farmers' intellectual property rights of various sorts may develop. Yet this possibility is far from inevitable. As Shiva notes:

> The dominant paradigm of intellectual property rights (IPRs) only protects innovation in the industrialized West. Centuries of innovation in the Third World are totally devalued to give monopoly rights to plant material to transnational corporations who make minor modifications compared to the evolutionary changes that nature and Third World farmers have made. IPRs thus place the contribution of seed companies over and above the intellectual contribution of generations of Third World farmers [active] for over ten thousand years in the areas of conservation, breeding, domestication and development of plant and animal genetic resources.
>
> Two biases are inherent in this argument. One, that the labour of Third World farmers has no value, while labour of western scientists adds value. Secondly, that value is a measure only in the market.
>
> (1994, 15)[11]

In fact, suggests Pat Mooney (quoted in Shiva 1994, 15), "The argument that intellectual property is only recognizable when performed in laboratories with white lab coats is fundamentally a racist view of scientific development."[12]

Scientists at IRRI and elsewhere—and particularly scientists who are also host country nationals—would no doubt be shocked to be labeled racist. However, there is no doubt truth to the idea that the labor of scientists is often valued (by politicians, bureaucrats, funding agencies, scientists, and sometimes even farmers themselves) more highly than farmer experimentation. While this may or may not be racist it is certainly *scientistic*—that is, this view privileges a certain kind of scientific knowledge over other ways of knowing, including farmer experimentation, which is labeled unreliable, idiosyncratic, or simply "not real science."[13]

The problem is not so much what crop science professionals explore as what they ignore. Vandana Shiva refers to professional crop development as a form of "epistemological violence." In professional rice science, as in other scientific disciplines, says Shiva:

> The object of study is arbitrarily isolated from its natural surroundings, from its relationship with other objects and observer(s). The context (the value framework) so provided determines what properties are perceived in nature, and leads to a particular set of beliefs about nature.
>
> There is threefold exclusion in this methodology: (i) ontological, in that other properties are not taken note of; (ii) epistemological, in which other ways of perceiving and knowing are not recognized; and (iii) sociological, in that the non-expert is deprived of the right both of access to knowledge and of judging the claims of knowledge.
>
> All this is the stuff of politics, not science. Picking *one* group of people (the specialists), who adopt *one* way of knowing the physical world (the reductionist), to find one set of properties in nature (the reductionist/mechanistic), is a political, not a scientific, act.
>
> (1988, 236)

In the case of in situ farmer experimentation, both peasant farmers *and* the object of their attention, rice seed, are passed through the filter of scientism and are thus denatured and misunderstood by professional scientists.

In the case of "modern" rice development, what has been excluded from the equation? Through a narrow focus on yields, other properties present in indigenous rice-growing systems are overlooked (Shiva's "ontological exclusion")—for instance, the self-contained sustainability of previous farming systems; the low cash costs of such systems; the disease resistance inherent in genetically diverse, rather than monogenetic, crop; and even the *taste* of the crop.

Alternative, "unscientific" forms of knowledge—other ways of seeing and knowing—are devalued (Shiva's "epistemological exclusion") by professional crop breeders. When farmer experimentation is defined as "haphazard," "local," or "idiosyncratic"—not the "universal" science supposedly produced by IRRI and like institutions—it can easily be dismissed by scientific "experts."

Finally, the scientific method of professional crop development often results in the exclusion of nonexperts from the rice development process (Shiva's "sociological exclusion"). The idea of farmer experimentation with rice becomes a joke, a contradiction of terms, a manifest impossibility.

In this case, at least two strategic avenues of response are available to farmers and their supporters. First, proponents may emphasize the practical, real world benefits of farmer experimentation—discussions of "indigenous knowledge" are one obvious way in which farmer knowledge is valorized by social scientists and farmers alike. Second, farmers may in fact become more like "real scientists" (as conceived by the scientific and political establishment); they may in a sense *take advantage of scientism* to assert that they, too, are able to systematically and "scientifically" develop new rice varieties and rice-growing technologies.

But is this second response even a remote possibility? Can "peasant science" (as I have taken to calling this strategy) be anything more than an oxymoron in the minds of policymakers and others in positions of power? More generally, if increased in situ preservation of rice germplasm is to be more than a pipe dream, exactly how is this to be accomplished? As Jackson notes:

> There has been only limited study of on-farm conservation systems worldwide and many questions remain to be answered about how on-farm conservation can become a viable method for genetic conservation. There is little understanding of the socioeconomic and genetic aspects of on-farm conservation. In what way do varieties change over time? Do farmers conserve varieties or conserve traits, such as aroma, plant architecture, or disease resistance, for example? What is the importance of seed exchange among farmers for enriching their germplasm? Why, for example, do some farmers continue to grow their local varieties while others abandon them in favor of improved varieties? What is the degree of outcrossing between varieties in farmers' fields?
>
> (1995, 271)

Obviously there is much research to be done, and we can only begin to touch on these questions here. However, a case study can provide some illumination. In fact, on-site preservation and expansion of rice genetic resources is taking place today in the Philippines, under the auspices of a nationwide network of affiliated farmers' organizations known as MASIPAG.[14] The MASIPAG network provides a case study of farmers who, with some initial help from sympathetic academics, have become what amounts to professional plant breeders, "peasant scientists" if you will, in an effort to exercise more control over rice production. The MASIPAG case illuminates and challenges the supposed division between allegedly "backward" or "traditional" farmers, on one hand, and "modern" and "scientific" researchers, on the other. It also suggests what might be accomplished by farmers interested in on-farm preservation of indigenous landraces *and* the invention of *new* rice varieties.

A "Peasant Science" Model of Biodiversity Maintenance

> The greatest service a citizen can do for his country is to add a new crop for his countrymen.
>
> —Thomas Jefferson

In the beginning (or for at least several thousand years prior to the 1960s), farmers controlled the raw materials of rice improvement: tens of thousands of individual seed varieties laboriously and continuously selected by farmers to best fit a wide range of environmental conditions. Thus, it is ironic that many crop scientists today apparently believe that the most useful place for rice seeds is in a giant refrigerator, not a farmer's field. In a sense, the scientific community has said, "Sure, these varieties all used to exist in farmers' fields but what *good* were they there?" From a crop science reference point, those strains were less than efficiently used: these landraces were unrecorded and uncataloged, virtually unknown outside a particular locality; hybridizations were haphazard and occurred only at the whim of nature; farmers *selected* what they considered to be promising varieties from natural hybridizations but didn't know how to make the purposeful artificial crosses that would have made the development of new seed types far faster and more efficient; and so on.

However, over more than a decade the MASIPAG network has begun to change the terms of this equation. Today thousands of farmers in scores of affiliated Philippine peasant organizations produce and maintain hundreds, perhaps thousands, of discrete rice varieties in ways that are recognizably, from the twin perspectives of crop science and agricultural politics, *scientific*.

The genesis of MASIPAG is a complex story, one I have recounted at length elsewhere (Frossard 1994). However, in brief, the organization is a product of the coming together of three particular groups, each of which was politically active in the waning years of the Ferdinand Marcos regime. The first of these, the Agency for Community Educational Services Foundation, or ACES, was and remains a nongovernmental organization involved in community political organization and rural economic development. It worked in the 1970s with farmers from the Central Luzon "rice basket" province of Nueva Ecija (where the first MASIPAG experimental farm would later be established).

By July 1985, ACES was able to gather together the second part of what would become MASIPAG—diverse farmers' organizations outraged at the steadily rising costs (both economic and ecological) of chemical- and cash-intensive Green Revolution farming. A farmers' convocation, sponsored by ACES, was held in Los Baños, Laguna—the center of rice technology development in the Philippines and home to both IRRI and to the University of the Philippines' main agricultural campus.

Farmers from Luzon, the Visayas, and Mindanao gathered in Los Baños to explore and, they hoped, find alternatives to their deteriorating prospects on the farm. They met with government representatives and IRRI scientists in sometimes acrimonious sessions. These tense confrontations embodied a subtext of opposition to the Marcos regime, as Marcos to some degree had staked the reputation of his government on the production of continuous rice surpluses using Green Revolution technology.

That subtext was readily apparent to the third group to join in the formation of MASIPAG, disaffected anti-Marcos scientists and university academics belonging to an opposition body they called the Multisectoral Forum, or MSF, ("multisectoral" in that not only were the academics and scientists of the University of the Philippines, Los Baños (UPLB), involved but students, and campus staff as well). For years, the MSF had been critical of the effects of Green Revolution technologies on rural smallholders. At the farmers' convocation, MSF members joined in the farmers' discussions and offered their services. But what exactly could academics—agronomists, soils specialists, and even a senior statistician—do for farmers? "Teach us to do what IRRI does," they replied. "Teach us to hybridize rice."[15]

In an incredible intuitive leap, farmers, academics, and community organizers realized that a possible solution to the increasingly disadvantageous terms under which small farmers labored could come through *greater control of the rice production apparatus itself.*[16] If pesticide prices were rising too fast, why not develop new rice varieties (or new multivariety cropping systems) that minimized the need for them? If fertilizer was becoming too costly—and if larger and larger amounts were now needed to maintain constant yields, a problem MASIPAG peasant farmers had identified a decade before IRRI (see Cassman et al. 1995; and Cassman and Pingali 1995a, 1995b)—then why not develop rice varieties less dependent on chemical fertilizers, or new ways to fertilize rice without purchasing chemicals?

It is interesting to note that these supposedly "traditional" farmers not only evidenced a good deal of faith in scientific rice production but also in their ability to carry out such experimentation (with some initial help from the academics). These farmers rejected not science but the products of an institution (IRRI) that for too long had been *conflated* with science. They resolved not to abandon science but to improve on scientific crop development by remaking it meet to their needs.

Thus, MASIPAG was born. By 1986, land for an experimental farm had been volunteered in Nueva Ecija Province, northeast of Manila. Training sessions in hybridization and crop-breeding strategies were conducted by specialists from MSF and ACES. Eventually, farmers produced their first hybridizations.

Since its inception, MASIPAG has institutionalized several goals (Salazar 1992, 25).

1. To develop new crop varieties that require few inputs, but (because they are particularly well-suited to local conditions) give reasonable yields.
2. To encourage farmer participation in rice breeding, nursery management, and the evaluation and selection of promising varieties according to farmers' own preferences.
3. To gain control over promising seed lines for planting and use as breeding stock.
4. To establish farmer-run seed banks to help resist genetic erosion.
5. To simplify the process of selection and dissemination of promising new varieties.

In many ways, as I will show below, MASIPAG has been an unqualified success. However, this is not to say that it did not face great difficulties (some of which the organization contends with today). For instance, there was an obvious logistical question for the "peasant scientists" to answer. Where would they find the first seeds with which to begin their trials? As it happened, Nueva Ecija was one of the first places in the world to receive IRRI seeds. From the first plantings of IR8 in the 1960s through decades of intensive government proselytizing, this province has been perhaps the most extensively "Green Revolution-ized" place on Earth. By 1986, indigenous landraces were few and quite far between in the province.[17]

Dr. Angelina Briones, a longtime member of the MSF and the UPLB Agronomy Department, recalls how some of the first MASIPAG seed stocks were acquired. When invited to speak or present a program outside the campus, academics working with the MASIPAG farmers would ask for their "payment" in a particular form. "When they received an invitation," said Briones, "they would write back, 'We are accepting this invitation and will be coming over, but please ask your farmers to bring a *dakot*, a handful, of the traditional varieties of seeds they have'" (quoted in Frossard 1994).

To build their germplasm collection, MASIPAG participants also acquired seeds from relatives in distant provinces, from other disaffected farmers throughout the country, and from the few remote areas of the Philippines where the Green Revolution was still just an idea. Even the omnipresent IRRI-type seeds were fair game for collectors (although MASIPAG farmers, for reasons that may have included pure nationalism, professed a preference for high-yield strains developed by Filipino scientists at the University of the Philippines, over varieties developed by expatriates at IRRI).[18]

In addition, MASIPAG faced other difficulties because of government infrastructure or agricultural policies. When situating its first experimental farm, the organization was forced to choose a site in Nueva Ecija, in part because the land had an existing deepwater well. Because government-built irrigation systems are run on a government-mandated timetable, only particular kinds of short-duration, fast-maturing (modern) rice can be grown in them. With its own well, MASIPAG was able to experiment with indigenous rices that grow more slowly.[19]

Also difficult for MASIPAG were government agricultural policies regarding bank loans. In part to minimize risk and maximize what it considered to be modern agricultural practices, government regulations often specified the particular rices to be grown, the pesticides to be applied, and the fertilizers to be added if farmers were to be eligible to receive crop loans.[20]

Over time, farmers hope, as MASIPAG grows more able to defend its seeds in *scientific* terms, the organization may also be able to assert that its own seeds are as appropriate and low risk as any high-yield varieties developed elsewhere—and thus as worthy of crop loans or insurance as other, government approved strains. This, however, must remain a long-term goal of the organization.

Despite these difficulties, MASIPAG has been able to maintain and multiply hundreds of promising rice varieties, keeping careful records of crosses. In its first year of operation, samples of 210 rice seed types—127 "traditional" varieties and 83 "improved" strains—were acquired from farmers' organizations and sympathetic individuals throughout the Philippines. In three years, one hundred hybridizations were made, with indigenous varieties strongly preferred by farmers for breeding purposes (Salazar 1992).

Because hybridization is "only 10 percent of the job," in the words of one participant, great attention is paid to the cultivation of test plots, elaborate record keeping, and eventual selection of promising types. A farmer "seed selection board" evaluates crosses for standard criteria such as disease resistance, productivity under a low-fertilizer regime, and even *good taste*.[21]

To maintain genetic diversity among the laboriously assembled collection, this MASIPAG station—like the many others that followed throughout the country—shared new seeds with farmers elsewhere, effectively testing these seeds under various soil, pest, and climatic conditions. By 1992, 40,000 kg of seed from thirty-four selected cultivars developed at several sites had been distributed in nineteen Philippine provinces (Salazar 1992).

As one MASIPAG experimental station mushroomed to thirty in 1992 and several hundred today (an exact count is nearly impossible because of the meteoric growth of the organization), affiliated farmers have produced, through standard scientific methods, hundreds of locally appropriate new varieties. The organization's scientists play a much less central role today, as farmers have mastered many of the basic techniques of scientific rice breeding and teach them to others directly. Through this widespread farmer to farmer contact, MASIPAG's "peasant scientists" not only preserve but *extend* rice genetic diversity in other sites throughout the country.

Conclusion

Willie Sutton robbed banks because, as he so logically pointed out, "That's where the money is." In at least the medium term, farmers and scientists will continue to depend on cold storage gene banks because that's where the rice is. Tens of thousands of varieties simply exist nowhere else. But, as I have argued, there are good reasons to encourage the propagation of a wide variety of rice genotypes in the field and at least one good model to show how that might be done in practice (and sooner rather than later).

But will farmers ultimately be an important factor in seed production and biodiversity maintenance? The Green Revolution apparatus certainly has many potential critiques at its disposal to denigrate such farmer experimentation. This is not to suggest the existence of a sinister conspiracy waged against farmers, only that the "common sense" of institutional crop science alone is enough to make peasant science seem absurd and unreal to professional scientists.

One such critique may lie in the *devaluation* of peasant science through the seemingly plausible claim that "true" science is, by its nature, an activity requiring extensive training unavailable to farmers—that any such science performed by peasants would be a relatively elementary science at best, hardly worthy of the name.[22] Or scientists, politicians, and other development industry players may simply claim that farmers *cannot* hybridize their own varieties, that such activities are beyond their knowledge.[23]

Certainly, farmers without access to the raw material of hybridization—a wide variety of rice genotypes and germplasm—would find it hard to refute these claims. This is the essential "Catch-22" of peasant rice farming today: If peasant farmers are to be seen as peasant *scientists* (or at least something more than passive recipients of institutional rice technology), they must first regain access to some of the indigenous seed germplasm now held in cold storage by centralized institutions. But to gain access to these carefully preserved seeds they must first be something other than peasant farmers.[24]

This is the danger—and the opportunity—that peasant science presents for institutional crop development. If MASIPAG can successfully redefine the supposedly traditional peasant farmer as (at least potentially) a modern peasant scientist, this opens up new opportunities for political assertion by the farmers—"regenerating people's space," in the words of Gustavo Esteva (1987)—and casts doubt on the supposed universality of work done at crop research centers. But let us not forget that MASIPAG also provides an opportunity for rice research institutions to use their seed banks in a more productive way—in concert and partnership with farmers—to increase biodiversity in the field rather than in the freezer.

Experiments in new forms of crop production such as peasant science arguably have the potential to advance rice development in ways that current Green Revolution modes cannot. Whether we as a species—a very *political* species with myriad competing interests—are able to take advantage of the inherent benefits of MASIPAG-style crop development and biodiversity preservation remains to be seen.

Notes:

1 Jackson (1995, 267) notes that while "the number of varieties of *O. sativa* is impossible to estimate ... claims of more than 140,000 have been made."

2 The institute's annual operating expenses (including genome storage) hover around U.S. $30 million. Similarly, smaller, international crop research centers based in North and South America, Europe, Africa, and Asia study and preserve other regionally important crops—from wheat, corn, and potatoes to cassava, chickpeas, sorghum, millet, yams, plantains, and a dozen others. The Consultative Group on International Agricultural Research (CGIAR), the umbrella body for IRRI and seventeen other such institutes worldwide, involves more than 1,000 scientists of sixty nationalities working with local specialists on food crop, small animal, and fish development. About 350 of these scientists, including 100 expatriates, work at IRRI (CGIAR 1993, 1996).

3 Although they are mentioned less frequently by Philippines farmers, who tend to see IRRI as the nexus of their troubles, other organizations also had a hand in this process. Donors include the Ford and Rockefeller foundations, national governments more interested in feeding restive urban populations than biodiversity, and transnational corporations that stand to profit by their control (through patenting) of seed germplasm.

4 The mechanism by means of which the scientific crop development industry gained hegemony over rice seed stocks is documented in Anderson, Levy, and Morrison 1991; in Fowler and Mooney 1990; and Ernest Feder's polemic *Perverse Development* (1983). Vandana Shiva (1991) holds a similarly jaundiced view of scientific control of germplasm through the modernist mechanism of the Green Revolution.

5 Within five years of its introduction, 50 percent of the rice grown in the Philippines was of the so-called modern varieties (Herdt and Capule 1983). By 1990, virtually all lowland Philippine rice came from seeds "improved" by scientists (IRRI 1992). As Thomas Hargrove has noted anecdotally, IR8 was known in Vietnam early on as "Honda" rice, since one or two harvests provided enough surplus rice to buy a motorcycle. Hargrove, later a public relations specialist for IRRI, was a military agricultural extensionist during the Vietnam War. He discovered many years after the war that he had been spared assassination at the hands of local Vietcong because he provided the highly coveted miracle rice seed to the area (Hargrove 1994). IR8 was nevertheless not particularly tasty to consumers and not particularly disease resistant; it was replaced with newer miracle varieties after a few years.

6 With depressed farm-gate prices and higher costs for petroleum-based chemical fertilizers and pesticides, high-yield (high-input) rice is no longer as attractive a proposition to farmers as it was in the 1960s. Unfortunately, farmers now typically find that they can not return to the

lower-yield, but much lower-cost, varieties they once grew, as they no longer exist in the field. Those who can return sometimes do so.

Soetomo (1992) reports a case in Indonesia in which net income to farmers growing a local landrace called Rojolele was 17 percent greater than with the high-yield IR64. Although the latter produced half again more rice per hectare, it also required twice the fertilizer, more weeding, and more water, thus making it less profitable for farmers in the end.

In the Philippine province of Cotabato in Mindanao, some farmers have turned to a farmer-discovered variety called Bordagol (meaning "short, solid, strong") which yields 20 percent less than Green Revolution varieties but requires only half the usual input of chemical fertilizers and grows so vigorously that weeds (and thus herbicide use) are minimized (Salazar 1992).

In another instance, explored later in this chapter, a group of Philippine farmers was unable to return to older varieties that had disappeared and so began an intensive self-education and rice-breeding program to develop new varieties appropriate to changing circumstances. Interestingly, unlike the previous examples, these farmers report no appreciable drop in yields with their new techniques.

7 It may be noted that international farmer-to-farmer seed exchanges have the potential to provide equal germplasm security.

8 In fact, the question remains whether farmers can continue to sustain these kinds of increases *indefinitely*. When Norman Borlaug, the architect of the Green Revolution, was awarded the Nobel Peace Prize for his high-yielding strains, he advocated not elation but caution. The advances of the Green Revolution, he said, were not the *solution* to humanity's food woes but merely a "breathing space" during which population growth could be brought under control. Chandler recognized then what few in positions of power seem willing to face: the ultimate source of the problems enumerated above is not insufficient yields but, *population growth*. The official neglect of overpopulation in most countries is unsurprising, however, as technological solutions to problems tend to be easier to "sell" than political solutions; increasing yields is a far more politically palatable alternative than serious population control measures in almost any country.

9 Brush (1992d, 145) maintains that the evidence so far shows "no general pattern of increased instability" in Asian rice or wheat yields due to the Green Revolution. However, Brush's numbers tell us only that a disastrous monocrop failure has only infrequently occurred to date, not whether it is now more likely to occur than before. He also reports on a group of Andean potato farmers who have maintained a high degree of genetic diversity in their crops while simultaneously planting "improved" varieties. While this is indeed a hopeful sign, I would suggest that such farmers are the exception rather than the rule. The farmers of MASIPAG are perhaps another such exception.

10 Unlike indigenous landraces, which are carefully selected by farmers and bred over many years for genetic stability, new high-yield rices tend

to "fall apart" genetically after one or a few cropping seasons, necessitating the purchase of new seeds. Not only does this add a new cash cost to the farm budget, but it further disempowers farmers, who are now crucially dependent on seed companies or government bureaucracies for their most basic necessity: seeds (see Shiva 1993a; Yapa 1993).

11 While Shiva paints a broad picture of the international IPR movement, Dove (1996) argues that only by understanding particular local arrangements—and the political-economic and development milieus within which they operate—can we successfully problematize the process of "intervention with intellectual property rights" and improve on that process. "The message, in short, is that there is an inherent bias in many existing structures of political power and resource use that is likely to cause most of the proposed uses of intellectual property rights to have either no impact or an impact that is the opposite of that intended" (61).

12 Mooney and Shiva are two high-profile figures in a much larger, international critique of bioprospecting—often referred to by opponents as *biopiracy*—and intellectual property rights laws as they now stand. Their critiques, often couched in terms of feminist theory, ethics, or philosophy, do not necessarily achieve wide currency in scientific or political areas but are well understood by farmers I interviewed.

13 As Cameron and Edge (1979) put it:

> In modern industrial societies, science plays a central and respected role. Scientific knowledge is taken to be the final word on, and is often called on to arbitrate over, "what is really the case." To "be scientific" is to earn admiration: it is to claim (and often to gain) credibility and superiority over unscientific alternatives—while to "be unscientific" is to be indefensibly wooly, vapid, old-fashioned, inefficient and generally unworthy of serious consideration. The community of scientists (which produces, legitimates and disseminates scientific knowledge) enjoys high status, substantial rewards and a position of power.

14 The word *masipag* itself means "industrious" in Tagalog. In this case, the term is an acronym for a phrase Mga Magsasaka at Siyentipiko Para sa Ikauunlad ng Agham Pang-Agrikultura, which roughly translates into English as the, "Farmer-Scientist Partnership for Agricultural Development."

15 In plant breeding, *selection*—the Asian farmer's usual means of improving plant varieties—involves identification of plants with particular desirable traits for further propagation over many generations. *Artificial hybridization*—hybridization performed by humans, rather than through natural means—is a scientific technique that involves the emasculation of a plant and its fertilization with the pollen of another. The purposeful mingling of genes from different plant varieties allows plant breeders to produce seeds that combine several desirable characteristics in one plant. Because this process often introduces undesirable genes as well, several generations of selection often follow the initial hybridization to minimize unwanted characteristics and maximize the desired ones. Many MASIPAG

farmers are well trained in hybridization techniques, performed with easily available tools: a small pair of scissors, a pair of tweezers (steel or bamboo), and a piece of cardboard.

16 How these farmers managed this leap of imagination is a fascinating—if perhaps ultimately unanswerable—question. Certainly, many other Philippine farmers in the mid-1980s were vocal in their demands for government subsidies to combat the highest fertilizer prices (and the lowest farmer monetary returns) up to that time. Why were the farmers who would later help form MASIPAG any different? Why did they believe they could take a quantum leap beyond high fertilizer prices toward a fundamental reordering of the rice-growing apparatus in which they labored? Part of the answer no doubt lies in the participatory, "bottom-up" community organization techniques employed by the ACES staff. Rather than telling farmers what their problems were and how they would be solved (a typical Philippine government approach), ACES asked the farmers to imagine a better life and offered to help them achieve the goals most of interest—ultimately, goals such as farmer production of locally appropriate seed varieties (Frossard 1994).

Too, we should perhaps credit ACES staffers and MSF crop scientists for their own leaps of imagination. By their own accounts (Soliman 1989, 77), it was difficult for them to rethink one particular idea so fundamental to their own training: that scientific knowledge and technology are fundamentally "beyond" the understanding of nonscientists. Ultimately, the stifling paternalism of the Marcos regime, against which they daily spoke out, may have helped these nationalist scientists surpass the bounds of their own paternalistic attitudes.

17 Indeed, in Nueva Ecija province, an early target of the Green Revolution and the founding place of the MASIPAG organization, the new IRRI varieties covered 99 percent of all rice lands by 1978–79—the highest rate in Central Luzon and twenty-five percentage points higher than the Philippine average at that time (Kerkvliet 1990, table B2).

18 Despite vague statements from MASIPAG members that UPLB plant scientists "understand better what farmers want," I was unable to pinpoint concrete physiological differences between the UPLB and IRRI high-yield rice varieties available to farmers. Undoubtedly, this is true in part because farmers have a keener eye than I for what constitutes a better rice plant. But given the organization's nationalist rhetoric, and given the many statements by MASIPAG members over the years about their preference for Filipino—rather than foreign—control of seed production, such a motivation cannot be ruled out.

19 If the products of the MASIPAG experimental farm are to be used later in these government-controlled irrigation systems, then quick maturation becomes a desirable quality—one that the MASIPAG "peasant scientists" have successfully incorporated into many of their new varieties.

20 As Goodell reported, Philippine farmers have often been forced through various government and bank edicts to adhere to very particular—and

not always locally appropriate—cropping schemes prescribed by government "experts." Failure to do so could result in denial of crop loans (Goodell 1984). Admittedly, however, in the tradition of James Scott's (1985) "everyday forms of peasant resistance," farmers devised various ways to circumvent at least some of these restrictions. One way was simply to sell unwanted inputs.

21 Although it was certainly productive, IR8 had very poor consumption qualities: It didn't taste very good and it didn't last very long before souring (which meant it couldn't be cooked in the morning and held to feed unexpected guests, an important part of Filipino hospitality). Not until IR64 did MASIPAG farmers find an IRRI rice that tasted particularly fine.

22 Former director general Klaus Lampe wrote that to meet the needs of millions more people IRRI "must initiate and facilitate rice research within the national agricultural research system of its developing-country partners" and to do so "must remain the major custodian of the world's genetic resources in rice" (1995, 257). In terms of IRRI's research into new rice-growing technologies, farmers apparently have no role to play.

23 When asked about the possibility that farmers could hybridize their own high-yielding, sustainable rice varieties, virtually all crop scientists I have interviewed, at IRRI and elsewhere, denied that it could be done. Farmers, they said, are simply unable to exercise the precision of technique and analysis—principally the mechanics of hybridization and record-keeping—to do what professional crop scientists do every day. When told that I had witnessed these practices myself, the scientists were incredulous and asked me to describe the process of hybridization used by the farmers. When I did so (correctly it seemed), they often appeared stunned.

24 It is claimed by IRRI claims that its seeds are free and available to all. In practical terms, however, because maintenance and propagation of seeds are so expensive, and because only small quantities of each variety is available, only tiny quantities of a few varieties are likely to be released to farmers at any time (although national seed programs and similar institutions have much better luck in this regard). In fact, frustrated MASIPAG members—farmers and professional crop scientists alike—have abandoned their so far futile attempts to acquire indigenous seed varieties from IRRI.

Part III: Biodiversity in Traditional Agricultural Systems

5

Fragmentation of the Ifugao Agroecological Landscape

MARILIZA V. TICSAY

THE EXTENT OF THE ENVIRONMENTAL AND SOCIAL PROBLEMS that accompany deforestation in the Tropics has become a major concern all over the world. In discussing causes of deforestation and genetic erosion in countries with tropical rain forests, many planners and decision makers point the finger at upland farmers who practice "shifting cultivation" or "swidden farming."

Policy-related descriptions of shifting cultivation often highlight the purported damage brought about by the slashing and burning of vegetation, soil erosion, and downstream siltation and flooding (Bandy, Garrity, and Sanchez 1993). Far less attention had been paid to the positive impacts and/or contributions of swidden farming such as the enhancement of soil fertility and decreases in soil aciditiy by plant ashes (Lawrence, Peart, and Leighton 1998; Lawrence and Schlesinger 2001), creation of habitat for species adapted for early successional stages (Brosius 1990; Conklin 1959; Wharton 1968), and even the creation and maintenance of floral and faunal diversity. The same is true of the relationship between shifting cultivation and local biological diversity (Conklin 1967a, 1980), the coevolution of swidden farmers and their environment (Collins 1986; Fujisaka 1986), or the ecological advantages of swidden farming in tropical conditions (Fox et al. 2000; Kleinman, Pimentel, and Bryant 1995; Rambo 1984; Thrupp, Hecht, and Browder 1997). Equally important, the practices subsumed under the term *shifting cultivation* (e.g., use of fire, minimum tillage, and natural and/or enhanced fallow [see Conklin 1961]) are typically not sufficiently differentiated to make it possible to understand the positive as well as negative impacts of each one. The same holds true for differentiation between shifting cultivation and the other agricultural activities that typically accompany it within a larger, composite agricultural system (shifting cultivation is rarely the only agricultural activity practiced by an upland community [Conklin 1980; Dove 1993a; Mahmud 1992; Padoch, Harwell, and Susanto 1998; Pelzer 1978]).

The environmental problems attributed to shifting cultivation can often be traced to the practices of migrant farmers who follow commercial logging operations into the uplands and open swiddens in logged areas but who lack both the proper incentives and the necessary technological knowledge for sustainable swidden cultivation (Brookfield and Bryon 1993; Cleary and Eaton 1992). There is growing evidence that traditional upland peoples, who have acquired extensive knowledge over time about their local environments, possess sound management technologies for sustainable agricultural production and resource utilization (Conklin 1957, 1980; Olofson 1981; Dove 1985b; Pei 1991). This study builds on these earlier descriptions of indigenous, upland environmental knowledge in two ways. First, I explore the traditional subsistence production system in an ethnic minority upland community in the Philippines that is in transition. My goal is to identify indigenous mechanisms and other factors that enhance biodiversity conservation even under conditions of change. I focus less on the various elements that comprise the swidden system or any other subsystem alone and more on how swidden cultivation along with a number of other land use systems articulate with one another to make up a successful, composite system of upland agriculture.

The second objective of this study is to develop methodologies for correlating the productivity and sustainability of different land uses (and at different hierarchical levels—e.g., species, field, and landscape) with variations in biodiversity conservation. Given the rapidity with which the environment is sometimes degraded, biologists' traditional, multiyear studies of enclaves of natural vegetation seem increasingly inadequate. The value of the rapid methodologies examined here lies in their ability to identify surrogate measures that can help us to assess the state of the environment, and the factors that both threaten and support it, in a much shorter period of time.

The Study Area

The study was conducted in several *sitios* (hamlets) of two contiguous *barangays* (villages) of Haliap and Panubtuban in the newly established municipality of Asipolo (formerly southern Kiangan) in the province of Ifugao (see fig. 5.1). Ifugao is located in the Central Cordillera mountain range in north-central Philippines (lat. 17°N). Barangay Panubtuban lies south of Barangay Haliap and is accessible only through Barangay Haliap. The study site is generally referred to as Haliap-Panubtuban and succeeding discussions and descriptions apply to the composite study site unless specific reference is made to one *barangay* or the other. The study site was selected because it is inhabited by members of a traditional ethnic community, Ayangan-Ifugao, and because it represents the sort of traditional, upland environment and society in the Philippines that is currently undergoing rapid transformation. The most prolific early ethnographer of the region was Roy Franklin Barton (1919, 1922, 1938, 1946, 1955); the most recent is Harold C. Conklin (1967a, 1967b, 1968, 1980).

Haliap-Panubtuban lies at an elevation of 800 to 1,200 m above sea level and is about 335 km north of Manila. It has a total land area of 2,800 ha and 304 households with a total population of 1,628. Ninety-nine percent of the population is of the Ayangan-Ifugao subtribe. Haliap-Panubtuban has been the focus of development assistance from government agencies, nongovernmental organizations (NGOs), religious and military groups, and professional and student researchers, which may have influenced the community's pattern of resource use and management practices and thus the level and pattern of biodiversity over time.

Resource Base and Patterns of Resource Use

Eight categories of land use that vary according to vegetation cover, terrain, hydrology, and agronomic activity were identified in the Haliap-Panubtuban landscape.[1] Not every household has access to

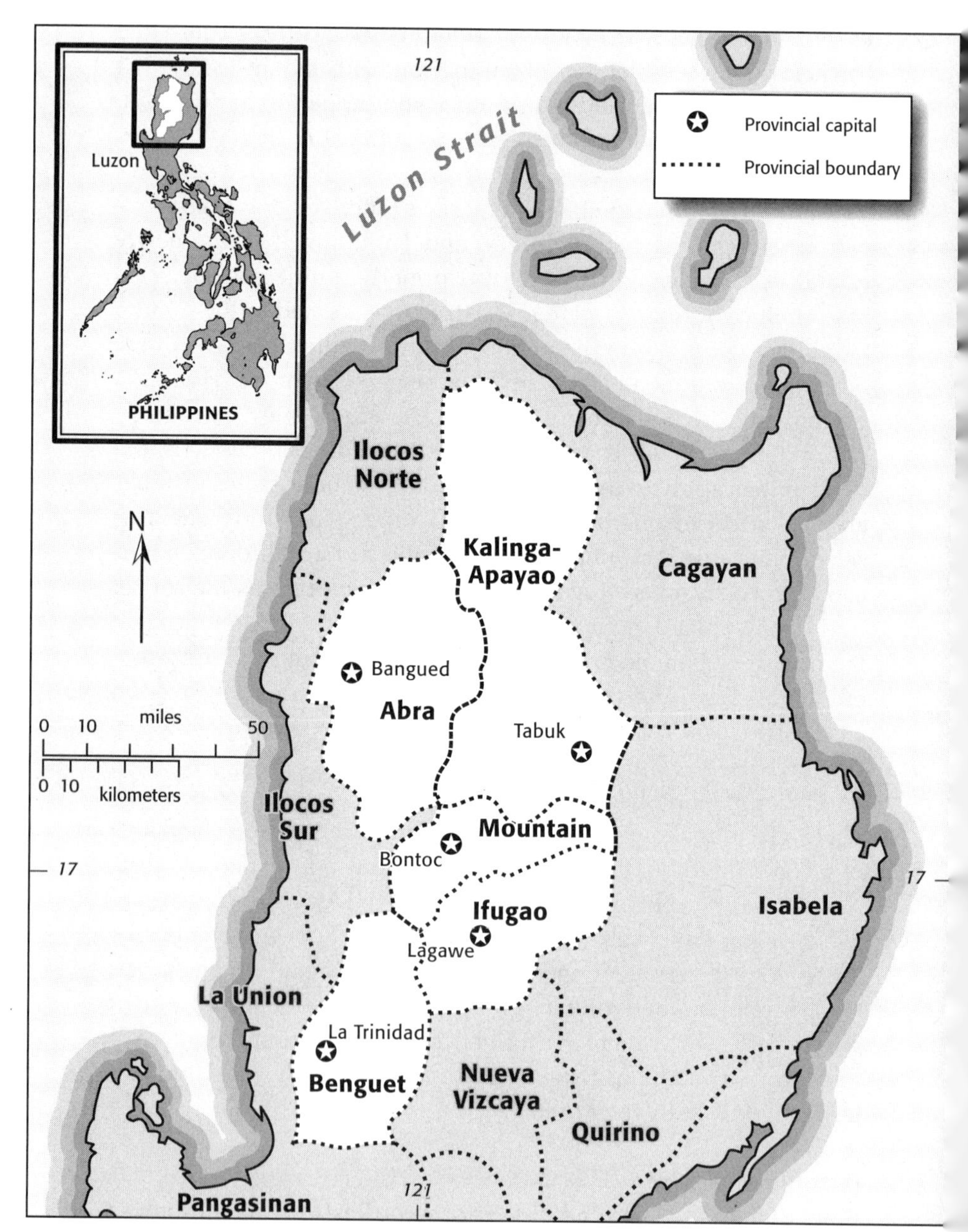

Figure 5.1 Geographic location of Ifugao, Philippines

and control of all resource bases. Two or more different resource bases also may overlap in the landscape. Each of these land use categories was analyzed for biodiversity rates using multiple methodologies (see below). Before presenting the methodologies used and the data found, I briefly describe each land use category.

THE COMMUNAL FORESTS ("ALA"). The (state-owned) communal forest areas are located in the highest elevations at the northern and eastern sides of both *barangays* and in the southern part of Barangay Panubtuban. Original forest cover can still be found there, and timber, fuelwood, and other forest products (e.g., honey and rattan) are also still freely obtained.

THE COMMUNAL GRASSLANDS ("KONGO"). The communal grasslands comprise 5 percent of the total land area of Haliap-Panubtuban, stretching mostly along the eastern slopes and partly along the southwestern portion of the study site. These areas are primarily used for pasturing goats, cows, and *carabaos* (water buffaloes).

WOODLOTS ("PINUSIU"). Woodlots are understory clearings within the margins of the forests. Customarily, forestlands may be claimed and cleared (as a declaration of ownership) by any pioneering farmer as long as the site has not been previously cleared and planted to coffee by someone else. If the woodlot is mostly planted to coffee, the site is called *nakopihan* (discussed below). Existing perennial vegetation in woodlots is managed to provide the household with a variety of products, including fuelwood, timber, fruit, and other resources. In some cases, the woodlot is planted to other tree crops (e.g., citrus, rattan, or bamboo). While these woodlots are privately owned and managed, just like the *ala*, any member of the community can freely collect fuelwood from "private" woodlots as long as timber species and coffee trees are not disturbed. Other forest products (e.g., honey) can only be harvested with permission from the "owner."

THE COFFEE PLANTATION ("NAKOPIHAN"). For many years, coffee was the only cash crop in the province. The Americans introduced it to Ifugao in the late 1800s, and the town of Kiangan became the coffee production center of the province (Manuta 1993). Coffee plantations spread throughout Haliap-Panubtuban as an economic response to the local people's need for a crop that can be bartered for trade goods. Importantly, biophysical conditions in the municipality of Asipolo in general and Haliap-Panubtuban in particular are suitable to meet the agronomic requirements of coffee (Barton 1922; Dixon 1990). The high demand for coffee in the early years further encouraged its production and resulted in the conversion of many woodlots to coffee plantations.

THE SWIDDEN ("INUMA-AN"). Swiddening is a year-round activity of the Ayangan of Haliap-Panubtuban, starting with site selection. The swidden is made in a forest patch, usually located on sloping lands, and composed of either new forests or old fallow fields, which is entirely cleared and burned for cultivating crops. Swidden produce is intended basically for the household and its livestock. Crops are chosen to satisfy these needs but with an eye toward their potential market value in case of surplus. Swidden crops vary from cereals and grains (e.g., rice and maize) to pulses, leafy vegetables, root crops (e.g., taro, sweet potato, and cassava), biennials (e.g., banana and papaya), and small trees (e.g., coffee and guava) grown in a three-year cycle.

Swiddens are cultivated for a number of years before they are permitted to fallow for three to seven years. When the swidden is being cropped, it is referred to as a *habal*. A household may maintain one or more habal at a time, which may be located contiguously or separately. Management practices include traditional crop associations, minimal tillage for soil conservation, and enhanced fallow.

THE RICE TERRACES ("PADJOW" OR "PAYOH"). Their wet rice terraces have made the Ifugao people world famous. The level terraces, carved out of steep slopes and reinforced with stone walls, maximize the use of land and water resources. The *padjow* or *payoh* is privately owned and primarily planted with traditional varieties of irrigated rice. Boundaries are carefully delineated and are distinctly marked with the red *dongla* plant *(Cordyline fruticosa)*. The *padjow* differs from the *habal* (swidden) by the presence of water and terracing. A *habal* is never terraced and is always dry and/or rainfed. A *padjow* is usually wet (flooded) and terraced. Small compost mounds called *po-o* are also formed in the pond field during land preparation, where some vegetables (e.g., onion, mustard, *petchay [Brassica napus])* are planted. A pit called a *dolog* (measuring approximately 1 m in diameter and 1 m deep, located in the middle of the pond field during the rice-growing stage) is used for fish culture for home consumption.

THE BEAN GARDEN ("GARDEN"). Vegetables have always been integral to the Ayangan production system. A variety of vegetable crops have always been planted in swiddens and home gardens for home consumption and the surplus production sold at local markets. It was not until the mid-1970s, however, that an additional land use referred to as *garden* became a popular agronomic activity in Haliap-Panubtuban, after the Department of Agriculture introduced a program promoting vegetable production (Bajracharya 1993). Before then, the Ayangan land use followed that of Conklin's description (1980).

A garden is a plot that is terraced if sloping but is preferably flat and usually converted from a swidden or dry rice terrace. It is intensively managed specifically for commercial production of vegetables, usually snap beans *(Phaseolus vulgaris* var. *burik)* and sometimes crucifers (e.g., cabbage, *petchay*, or mustard).

The recent trend toward the conversion of swiddens and dry terraces to bean gardens is not entirely due to market conditions. It can be attributed primarily to current environmental conditions. In the early 1980s, when regional climate changes were experienced (prolonged dry seasons resulting in drier rice terraces), farmers were forced to find an alternative crop. Initially, they experimented with snap beans on any available land—a small portion of the swidden, on some drier spots in terraced fields, and small plots in home gardens. The crop's short growing season (three months per crop) and a strong and growing market value attracted the farmers to appropriate additional land for bean gardening. Several farmers in Haliap have actually opted for bean gardening as their entire source of livelihood, converting dry terraces and old swidden plots into large bean gardens. Bajracharya (1993) reported that the majority of bean gardeners are young, single males who can devote more of their time to intensive cultivation of farmlands and can take the greater risk inherent in producing a single crop with a fluctuating market value.

Although gardens are cultivated mainly for the production of cash crops, traditional rice varieties and other vegetable crops for home consumption can sometimes can be found planted on the borders of garden plots. Some weeds or uncultivated species are also tolerated (e.g., *Amaranthus* spp., *Solanum nigrum*, and *Nasturtium indicum)* and harvested for home consumption.

THE HOME SETTLEMENTS AND THE HOME GARDENS ("FUBLOY" AND "NGILIG"). The *fubloy* is a cluster of three to ten houses that make up a home settlement. Within the home settlement, the home garden is an essential resource base for household subsistence. The home garden is defined locally as the area occupied by and surrounding the house. It is separated into two distinct parts: the *ketao*, the open courtyard in front of the house; and the *ngilig*, the backyard, where there are more trees and a stone-walled livestock pen *(fu-hod)* is also typically located. A fishpond, if present, will be located wherever the water supply is most abundant. There is no single Ayangan word for either "home garden" or "fishpond," but villagers usually refer to the former as the *ngilig*. Generally, it is planted with ornamental species making up hedgerows and living fences around the periphery of the front yard. Vegetable species are either planted intentionally or casually (with seeds tossed out of the windows) for home consumption. Fruit trees, particularly the *Areca* palm, are planted not only to provide fruits and nuts but most importantly to serve as living stakes or visible markers of landownership.

The home garden also serves as a seed bank or repository of planting material and a nursery for acquisitions from other resource bases (e.g., orchids from the forests or seeds of fruits from the market) and an experimental plot for new crop species. In effect, the home garden is an extension of the swidden for the production of food crops; an extension of the woodlot for fruit, fuelwood, and construction materials; a source of shade; and an expression of aesthetics.

Review of Factors That Influence Biodiversity

Species composition and diversity in vegetative cover can be directly linked to the nature of the resource base and the patterns of land use conversion, especially in the more intensified production of agricultural crops in the Haliap-Panubtuban landscape. In the following pages, I explore the specific biophysical, sociocultural, economic, and institutional factors that can potentially influence land use conversion and consequent changes in biodiversity levels in Haliap-Panubtuban.

SITE QUALITY. Species composition and distribution (i.e., biodiversity) can differ significantly, as they are influenced by the varying elevation and topography of the land. For example, certain plant species (e.g., *Pinus* spp.) naturally grow at high elevations (e.g., 1,000 m asl) in Barangay Haliap. Land use can vary considerably by elevation and topography. A biophysical slice of the Haliap-Panubtuban landscape shows that communal forests are found at the highest elevations, with the rice terraces at the lower elevations and on less steep slopes. Swiddens are usually located above the rice terraces on steeper slopes, while bean gardens are located on flat plains. Home settlements may vary slightly but are usually located at lower elevations with flatter topography.

Soil fertility and acidity have a profound influence on crop productivity, as both influence the type, species, and varieties of crops to be planted. In areas of high soil fertility such as in a newly opened swidden, highly valued crops such as upland rice are favored for the initial planting. Less demanding crops or crops tolerant of low nutrients, such as sweet potato and cassava, are planted as soil fertility decreases through years of cultivation. The decline in soil fertility as perceived by the farmer is the primary reason for permitting the farm plot to lie fallow. The stoniness of the site determines workability and consequently crop suitability. Furthermore, physical soil characteristics such as structure, texture, and porosity determine moisture absorption and water availability and retention. Most of Haliap-Panubtuban is steep and rocky with limestone outcrops and is suitable for root crops that require minimum tillage.

WEATHER AND CLIMATE. Rainfall and the prevailing temperature modify the vegetation of an area, as these greatly affect agricultural activities and cropping patterns. Successful introduction and domestication of new crops largely depend on climatic compatibility. In Haliap-Panubtuban, precipitation is a limiting factor. During the two-year duration of the study (July 1991 through August 1993), the climatic conditions in the area (and the entire country) were described by key informants as "not normal." There was a prolonged dry season in early 1992, resulting in crop failure of both irrigated and rainfed rice planted in October 1991 and February 1992. Furthermore, the normal April and May showers did not come, delaying swidden cropping scheduled in late May until early June. The rains that finally came were severe, with strong winds and torrential downpours that devastated a great deal of the northern parts of Luzon, including the province of Ifugao. The extended rainy season in 1993 set a record for the maximum number of typhoons in a single year in Philippine recorded history (Philippine Atmospheric, Geophysical, and Astronomical Services Administration Weather Report, Dec. 1993).

HYDROLOGY AND ACCESS TO WATER SUPPLIES. Water is extremely important for both domestic and agricultural uses, particularly for wet rice terraces. A regular water supply is not entirely dependent on rainfall but also on conditions of watershed vegetation. The deep humus layer of undisturbed forest soils reduces quick run-off of rain waters. The drying up of springs and general water shortages due to the aforementioned drought dried up the rice terraces, particularly the upper ones, which has been a primary factor in the conversion of pond fields to bean gardens.

CROP CHARACTERISTICS. The choice of crops is highly use-defined. Certain crops are more favored or valued due to their high productivity and low material (e.g., fertilizers) and labor requirements. Other important considerations are better taste (as in some traditional rice varieties), marketability (e.g., snap beans and coffee), multiple utility as feed or fodder (e.g., sweet potatoes and maize), and environmental influences of a protective cover crop (e.g., sweet potatoes). Certain crops can therefore serve as good indicators of site quality (e.g., graminoids vs. perennials) and farm family conditions such as market orientation (e.g., snap beans), tenurial security (e.g., Areca palms and perennials), cropping stage within the life of the plot (e.g., first vs. second crops and rice vs. vegetables), household preference, and other land use activities (e.g., rice vs. beans or fodder species). Specific plant indicators are enumerated below.

ACCESSIBILITY OF THE FARM. Distance of the farm from the farmer's home settlement is another important criterion in site selection. Most swiddens and pond fields are located within fifteen to thirty minutes' walking distance. Proximity to one's other plot(s) may also be important, as this gives easy access to a farmer who may be cultivating two or more swiddens in addition to a rice field and who has to reach his other home in time to prepare the family's meals. Most bean gardens are established near existing roads and trails to facilitate market transfers.

THE TRADITIONAL CROPPING SYSTEM. Although the high market value of beans has resulted in land-use conversion, it was also observed that many older farmers who opted for intensive bean cultivation for cash revert to rice farming as soon as irrigation water becomes available with the onset of the rainy season. Some bean gardeners also managed their gardens differently and had developed slightly different cropping schemes that put less pressure on the soil. According to key informants, planting the land with a crop other than rice (e.g., snap or mung beans) during its rest period helps to restore its fertility: the first crop, planted in January, is rice (a four-month crop); the second crop, planted in June, is beans (a three-month crop); and this may be followed by a third crop of beans planted in September. The land may be left to lie fallow until the year begins again with rice. The choice of beans as the alternative crop is influenced by the market. Before beans became popular, rice was normally planted in the terraces or swiddens in January, with an option for a second rice crop in June or July. The other option was to leave the terraces fallow for the rest of the year and concentrate household labor on swiddening and other nonfarm or off-farm activities such as weaving and basketry or wage labor, which are alternative sources of cash.

The cropping system in Haliap-Panubtuban is the result of the farmer's perception of household requirements not only for food but for social obligations, cultural rituals, and cash. The cropping system of the Ayangan of Haliap-Panubtuban is therefore "a mosaic of a variety of crops including rice, maize, legumes, sweet potato, banana, papaya, betel nut, coffee, pomelo, rattan, and other food and tree crops" (Guy 1995). As the cash and material needs of the household increase, modifications in the traditional resource management system are made to include more income-generating activities.

Animal production is an important component of many traditional agricultural systems, including that of the Ayangan-Ifugao. Not only does livestock provide draft power (e.g., *carabao* transport market produce and assist with logging) and food products (milk, eggs, meat, and hides), but certain animals are important in Ayangan rituals. Dowries *(fonong)*, thanksgiving feasts *(falaong)*, and some traditional rituals *(funinalishong)* require animal offerings to deities and spirits. Moreover, the Ayangan system of kinship is manifested in a ritual of meat sharing *(ipad)*. The importance of livestock in the Ayangan way of life is reflected by the crops cultivated on their farms (sweet potatoes, maize, cassava, and taro) not only for home consumption but, more importantly, for animal feed. The most common livestock raised in the area is the native pig, which is a requirement in many sociocultural and religious celebrations and rituals. Chickens and ducks are also used for ritual offerings and home consumption. Cattle and *carabaos* are employed as draft animals. Goats are becoming increasingly popular for meat, and dogs, which are the usual house pets, are also a delicacy in Ifugao cuisine (e.g., in *kinilaw*).

LABOR AVAILABILITY AND MATERIAL INPUTS. Although cash crop production is usually profitable, it is more expensive in terms of labor requirements. Aside from the expected increase in material inputs, such as inorganic fertilizers and pesticides, this increased demand for labor is a deciding factor in whether a household will convert from subsistence production in the swiddens and rice terraces to cash crop production in the bean or vegetable garden. In many cases, farmers, especially swiddeners, who are mostly women, do not convert or may have tried it once and reverted to swiddening due to the increase in labor inputs required for bean gardening.

For households that opted for bean gardening as the major farm activity, a significant drop in crop production for home consumption has been observed, resulting in a decrease in new forest openings for swidden. At the same time, land allotted for bean production becomes intensively and continuously cultivated with the use of chemical inputs. Furthermore, it was observed that available space is planted to an assortment of subsistence food crops for daily home consumption (e.g., onions, mustard, and eggplant), which would normally be planted in the swidden.

LAND TENURE. Manuta (1993), in his study of tenurial arrangements and resource use management in Barangay Haliap, stated that with the increase in population and the system of property transfers throughout the generations migration to other areas has been the village response as the topography of the area limits the expansion of agricultural land. Manuta's study also indicated that security of tenure and differential access to key resources have a direct bearing on how resources are used and managed. A farmer who has more landholdings and/or bigger parcels can afford a longer fallow period because of the availability of space for plot and/or subplot rotation. He or she can leave a plot to regenerate naturally or plant perennials and abandon it. The perennials usually consist of fruit and timber species that can be of use in the future.

THE TRADITIONAL IFUGAO SYSTEM OF PRIMOGENITURE OF INHERITANCE. This system prohibits fragmentation of the family property, especially land. The eldest child inherits all the land, and the younger children must find other employment, develop alternative sources of income (off-farm or nonfarm), or open new lands for themselves elsewhere. However, as tradition is increasingly being overlooked by succeeding generations, inheritance schemes are changing to allow younger children to inherit some land from their parents, depending on their total number of landholdings. As a result, members of the younger generation, even those who choose to find employment elsewhere, are able to come back to Haliap and venture into bean gardening. These modifications in inheritance still do not allow parcels to be fragmented. When the heir claims currently cultivated land at the time of marriage, parents open up new parcels, usually for swiddening, from landholdings that had been neglected, abandoned, fallowed, and allowed to revert to forest growth or from previously uncultivated forest. As there does not seem to be any more land available for pioneering in the communal forest of Haliap-Panubtuban, the elders are establishing swiddens in old woodlots that resemble secondary forest.

TRADITIONAL BELIEFS AND PERCEPTIONS. The traditional Ayangan believes in spirits (both "good" and "bad") called *finacheng* and *fifigiew* that inhabit "holy" trees. These trees are usually large and old, are covered with creeping vines and lianas, and grow near springs and rivers. Village folks believe that these spirits can cast misfortune and cause illness when disturbed and angry, and for this reason these holy trees are left undisturbed and in effect are "protected."

The tree species most revered as a holy tree is the *balete (Ficus benjamina)*, which is frequented by fireflies *(alitaptap)* at night. Another tree species identified as holy is the *fulala* tree *(Octomeles sumatrana)*. There are only two *fulala* trees in the area—one in Sitio Binablayan, Haliap, and the other in Sitio Nuntiguing, Panubtuban. Both trees, over 30 m high, are growing on the banks of stream tributaries of the Itum River, and are covered with a variety of mosses, ferns, orchids, and other epiphytes.

GOVERNMENT PROGRAMS. In 1980, the Department of Agriculture, in its bid to intensify agriculture through the introduction of new crops, launched a program promoting vegetable production called the Save the Terraces Program. The program works on the premise that bean gardens are a good alternative to rice cultivation in Ifugao, considering the positive conditions for gardening (i.e., temperate climatic conditions, sufficient water resources, and the amount of labor available).

Although the program was intended for the entire province of Ifugao, it was launched in the municipalities of Banaue and Asipolo. The residents of Asipolo (mostly Ayangan) were more receptive to vegetable gardening than those of Banaue (mostly Tuwali), who were reluctant to grow vegetables on the terraces in the belief that paddies and rice cultivation are inseparable and that draining the paddies for gardening would destroy the harmony between paddies and the ancestors who built them. Draining the terraces is considered taboo and, it is feared, may lead to disasters such as landslides.

Vegetable production is integral to the subsistence production system of the Ayangan of Asipolo, where the people are very keen on eating vegetables at every meal. This is very apparent in the higher occurrence of malnutrition in Banaue compared to Asipolo, according to a undated report by the United Nations Children's Emergency Fund (UNICEF), cited by De Boef (1990). Vegetables are often cultivated in areas too steep or dry for the construction of paddies and rice terraces, on terrace walls and/or dikes, and on compost mounds in rice paddies. Most often, flat to gently sloping grasslands or pasturelands are also transformed into vegetable gardens. The variety of vegetable crops planted in the gardens includes crucifers, cucurbits, green amaranth, pulses, and tuber/root crops. By the time the Department of Agriculture began promoting vegetable production, the people of Asipolo were already experienced in vegetable gardening and were ready to do so on a commercial scale. Soon Asipolo became a vegetable-producing area, with many farmers growing snap beans *(Phaseolus vulgaris)* in monoculture. The cultivation of snap beans and the conversion of paddies into bean gardens became even more attractive when the province began experiencing a prolonged drought, resulting in insufficient water for rice in the terraces and making them more suitable for vegetable cultivation.

EXTERNAL MARKET FORCES. Bajracharya (1993), in her study of the influences of commercial agriculture on Ifugao culture, stated that the main indicator of participation in the cash economy is the market-oriented production of agricultural crops. The production system in Haliap-Panubtuban has been observed to be a mixture of subsistence and market-oriented production, which is called "an economy in transition."

The major cash crops in the area are snap beans, coffee, and pomelo. The spread of coffee plantations in Ifugao was a response to the local need for a crop that can be bartered for other goods. The high demand for the crop in the early years further encouraged the production of coffee. The slump in coffee prices in more recent years, however, has discouraged many farmers, and as a result most coffee plantations have become overgrown with shrubs and vines. Many have been abandoned *(natagwan)* altogether or planted to other fruit tree crops such as pomelo, which also commands a high price.

As indicated earlier, the latest cash crop of Haliap-Panubtuban is snap beans *(Phaseolus vulgaris* var. *burik)*. The high market value of beans is the primary attraction for farmers to cultivate them on a commercial basis (Bajracharya 1993). It is cheaper and more convenient to buy rice for home consumption out of one's income from beans than to plant rice in the pond field. Bean gardening has become so attractive that many pond fields (especially in Haliap) have been intentionally drained for conversion to bean gardens.

Methodology

Multiple methodologies were used in data collection at different hierarchical levels, from the species to the landscape level, in order to assess conservation of biodiversity. The methods include a combination of the more traditional quantitative measures of biodiversity and recently developed qualitative methodologies, including, rapid appraisal techniques, interviews, and ethnobotanical studies. Baseline information on relevant biophysical, sociocultural, and economic aspects of the environment, as well as general trends of resource utilization in the study sites, was obtained from secondary data sources. Table 5.1 outlines each level of study and type of methodology employed to gather data. Below each methodology is described. Special attention to the types of methodology employed is warranted since this study relies on an unusual combination of social and biological information as well as qualitative and quantitative data. In the following section, the results of the study are presented.

Biodiversity Measurement and Analysis

QUANTITATIVE OR STANDARD MEASURES OF BIODIVERSITY. In order to establish a basis for comparison with biodiversity indices derived from other studies, standard indices such as the Shannon-Weiner index (H'), the Menhinick index of species richness (R), and the Pielou's evenness index (E) were computed for selected resource

Table 5.1 *Summary of methodologies used for biodiversity assessment and data analysis*

Scope and Objectives of the Study	Level of Analysis	Unit of Analysis	Measure of Diversity	Variables/Data Required	Data Acquisition/ Methodology/Technique Used
Resource survey and inventory	Plant community/ vegetation cover	Species, varieties, and cultigens	Species inventories; indicator species	Species: identity baseline biodiversity indices	Species inventory; rapid biodiversity appraisal; interviews
Identify and describe pattern of the traditional swidden-based production system of the Ayangan	Human community/_ agroecosystem level	Household or farm plot	Extent of monocropping; genetic erosion	Cropping calendar; crop association; economic plants; principal staples; cropping history; market orientation; standard of living; strength of local institutions	Review of secondary data; rapid rural appraisal; key informant interview; participant observation; life histories and case studies
Identify endogenous/exogenous factors that influence biodiversity management	Landscape level	Household strata; land uses	Species endemism, extinction, endangerment; landscape structure, functional attributes, patchiness	Cultural, sociopolitical, economic, and demographic profile (number of resource bases per household and household demand)	Interviews; life histories and case studies; statistical analysis of variance; correlation analysis
Rapid biodiversity appraisal methodology	Landscape level	Household, farm, resource base	Species count, surrogate parameters	Cultural, sociopolitical, economic, and demographic profile (number of resource bases per household and household demand)	Statistical analysis
Functional analysis of biodiversity	Species, resource base, landscape	Species	Landscape structure and functional attributes	Economic, sociocultural, ecological functions of biodiversity	Ethnobotanical survey; case studies

bases (holding size constant). These standard indices were then compared between and among the different resource bases. Sorensen's index of similarity (*IS*) and dissimilarity (*DS*), based on the presence (not the absence) of species, was likewise computed for the different resource bases.

RAPID BIODIVERSITY APPRAISAL METHODOLOGY. This approach to biodiversity assessment does not compute biodiversity indices but identifies factors or surrogate parameters that may indicate conditions of biodiversity (i.e., loss or maintenance). This methodology integrates biological and social information gathered from field observations and interviews at the study site. It involves only species identification and recording the total number of species found in a particular sampling area. It does not involve counting the number of individuals per species in a given sampling space to predict species distribution and abundance. Other procedures included in the rapid biodiversity appraisal methodology are (1) vegetation mapping and profiling, (2) "kitchen" and "market" surveys, and (3) the participatory transect walk.

SPECIES LISTING, VEGETATION MAPPING, AND PROFILING. Species listings were carried out in all 323 farm plots belonging to the ninety-three respondents of the survey. (Species listing is purely taxonomic in nature, with no frequency counts done for any of the species identified.) With the assistance of a local informant, a representative sampling plot (selected from the 323 visited) was measured and drawn to illustrate its exact configuration and dimensions. Plant crops were then identified and plotted to show species association. A vegetation profile was also drawn to show the vertical distribution of the different plant species from the ground to the top (canopy) story.

"KITCHEN" AND "MARKET" SURVEYS. Both the household kitchen and the village market are good places to observe the seasonality and abundance (reflective of crop failure or success for the season) of cultigens and noncultigens, the crop's market value (as reflected in price), and people's dietary preferences. Furthermore, the role of different crops in human nutrition can be deduced. This method involves listing, describing, and identifying the species (both plant and animal) found, cooked, and prepared for household consumption. The Asipolo market was visited weekly and the local store was visited daily. The kitchen survey was conducted daily.

PARTICIPATORY TRANSECT WALK. A transect constitutes a form of systematic sampling wherein the samples are arranged in a linear fashion. Transects are useful for recording changes in soil type, topography, elevation (Iskandar and Kotanegara 1993), and land use. I carried out an east-west transect across the landscape, and identified and recorded all plants along the way (see fig. 5.2). Frequency and occurrence are good indicators of the abundance and perhaps the functional roles of flora and fauna in the landscape. A local resident was asked to act as a guide (and interpreter, when needed) during the transect walk to facilitate the acquisition of local nomenclature as well as to gain insights into local and traditional uses of each species found.

BIOSOCIAL INDICATORS OF BIODIVERSITY. Just as species diversity can be measured quantitatively with the use of standard mathematical models, so can changes in the conditions of biodiversity be monitored and evaluated through the use of certain indicators. Note that indicators are not given; they are created. Because utility of a given species is user defined and highly subjective, these indicators may be highly site specific, meaning that an indicator for site A may not be an appropriate indicator for site B. Indicators can be identified from information collected in respondent interviews as well as from field observations during the rapid biodiversity appraisal. They point to particular land uses or other conditions (e.g., site quality or soil fertility) that may be associated with biodiversity levels. The presence (or absence) of particular flora or fauna and particular biophysical conditions or *barangay* rural activities is the first step toward identifing useful indicators.

DATA ANALYSIS. Analysis of data was carried out on the species, household, resource base, and landscape levels. At the species level of analysis, species richness in terms of taxonomic diversity was assessed by means of species count. A functional analysis (i.e.,

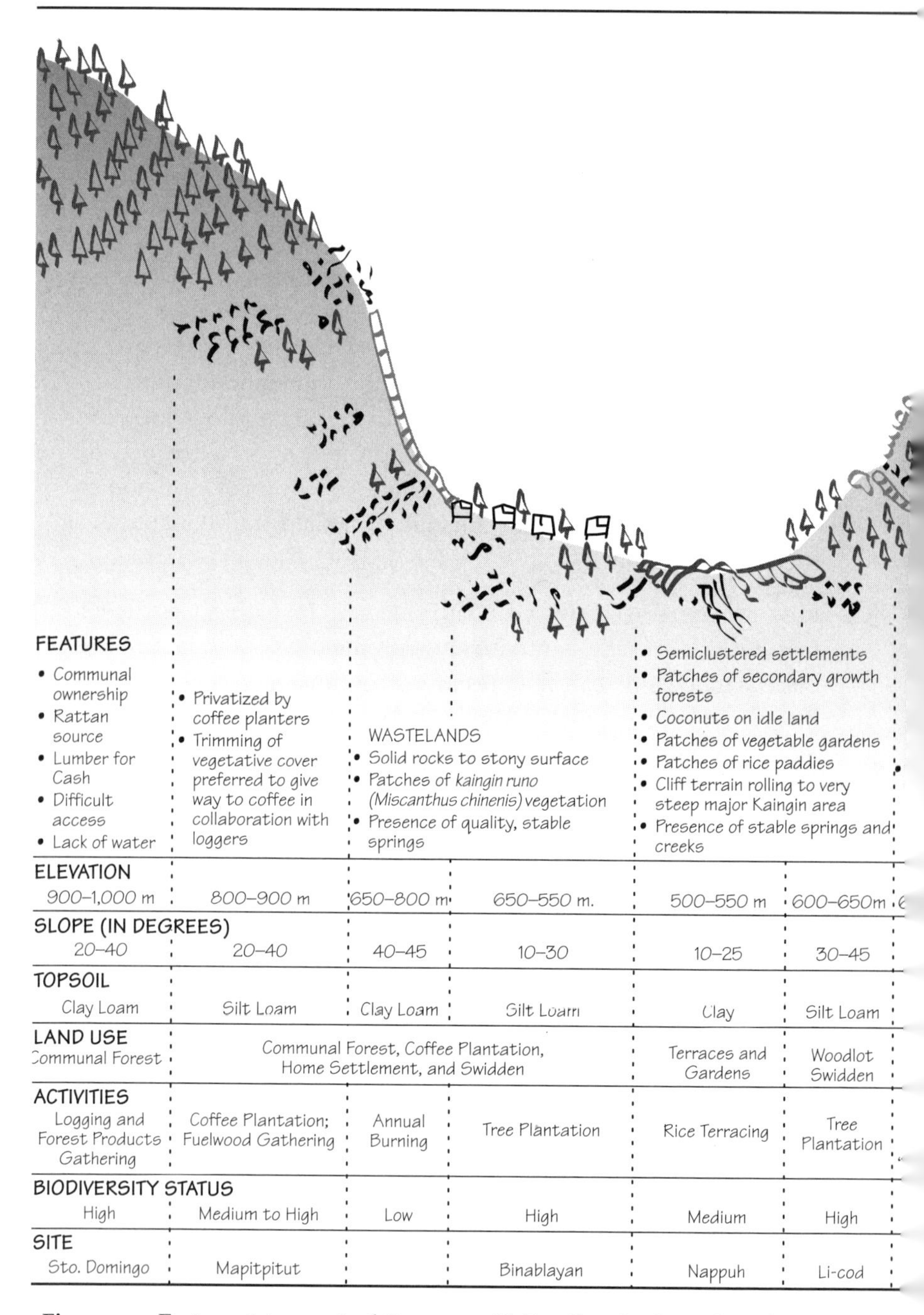

Figure 5.2 East-west transect of Barangay Haliap-Panubtuban, Asipolo, Ifugao

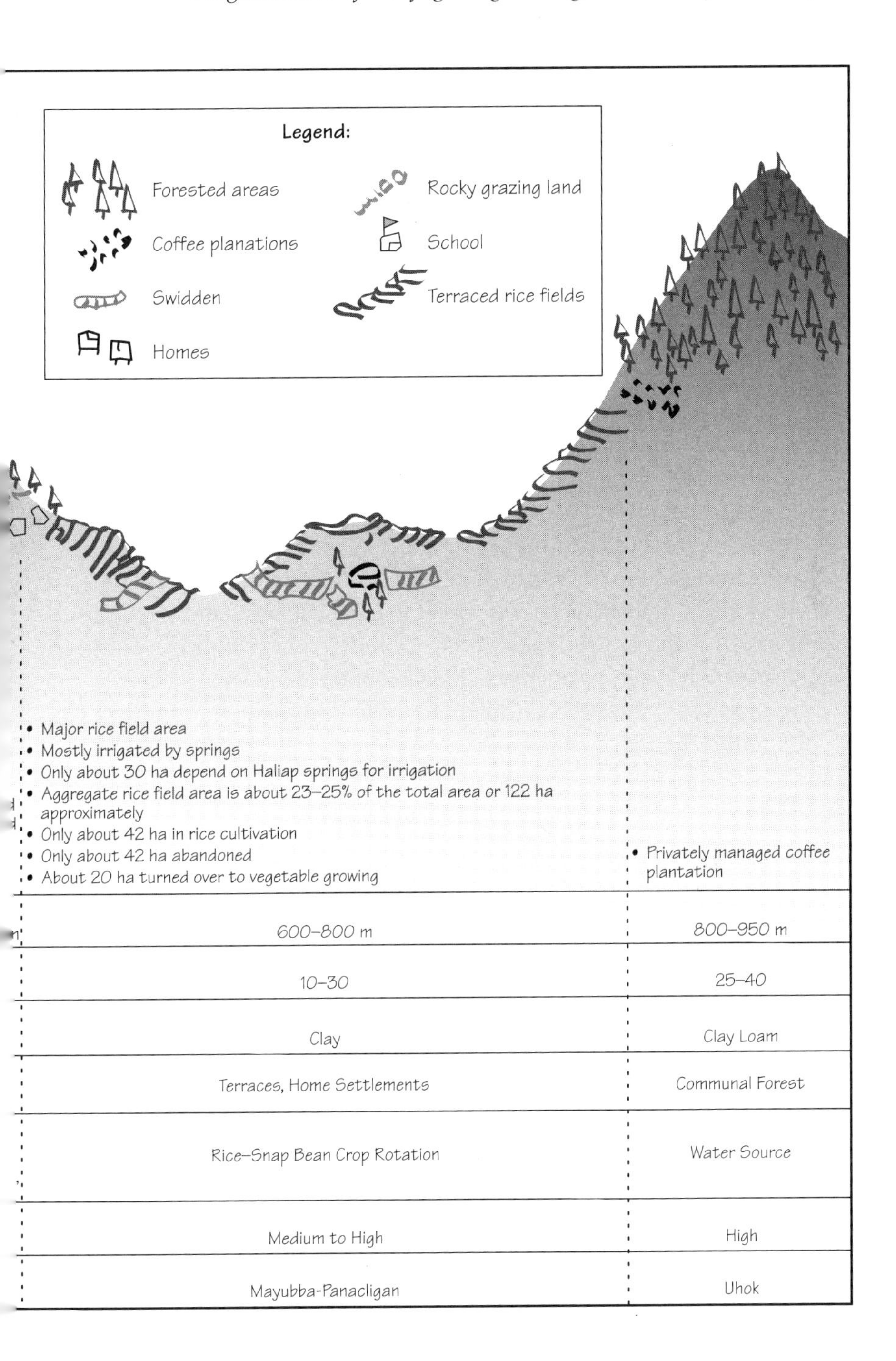
Legend:
Forested areas
Rocky grazing land
Coffee planations
School
Swidden
Terraced rice fields
Homes
• Major rice field area
• Mostly irrigated by springs
• Only about 30 ha depend on Haliap springs for irrigation
• Aggregate rice field area is about 23–25% of the total area or 122 ha approximately
• Only about 42 ha in rice cultivation
• Only about 42 ha abandoned
• About 20 ha turned over to vegetable growing
• Privately managed coffee plantation
600–800 m
800–950 m
10–30
25–40
Clay
Clay Loam
Terraces, Home Settlements
Communal Forest
Rice–Snap Bean Crop Rotation
Water Source
Medium to High
High
Mayubba-Panacligan
Uhok

economic, social, and ecological value or utility) was carried out using an ethnobotanical technique. At the household level of analysis, biodiversity as measured by species counts was assessed and compared among the different household respondents, and household variables explaining differences in biodiversity levels were identified (e.g., household demographics or multiplicity of resource bases).

At the resource base level of analysis, biodiversity as measured by species count was assessed and compared among different plots of the same resource base. Factors explaining the differences among plots were identified and assessed (e.g., site quality or tenurial status). In addition, plant indicators of biodiversity status were identified and assessed; functional analysis (i.e., economic, cultural, and ecological utility) of the resource base was also done in relation to its species content. At the landscape level of analysis, traditional Ayangan agricultural production was analyzed focusing on the factors affecting biodiversity levels at the level of each land use and the level of the entire landscape (e.g., indigenous soil and water conservation practices or crop management techniques). Analysis of Variance (ANOVA) and correlation analysis were carried out on the different factors.

Results and Discussion

Analysis of Biodiversity at the Household Level

Table 5.2 gives a summary of significant factors affecting biodiversity at the household level. Using the household as the basic management unit, the study identified and investigated factors that influence the utilization, management, and consequent conservation or loss of biodiversity in Haliap-Panubtuban. Household biodiversity (as measured by taxonomic diversity) is the total species count of all flora in all plots or resource bases managed by that household. From a list of households of each *barangay* obtained from the Barangay Council secretariats, ninety-three household respondents, or 31 percent of the population, were identified and interviewed regarding resource management. The factors that subsequently proved to influence resource management and the loss or maintenance of biodiversity included household demographic characteristics (age, gender, marital status), household size or number of consumers in the household, number of producers in the household (to indicate labor force), consumer-producer ratio of the household, sources of extra income (farm, nonfarm, off-farm), total number and total number of types of resource bases exploited by the household, inclusion of swiddens and home gardens in the resource bases of the household, vegetable sufficiency of the household, and livestock integration.

Statistical analysis shows that the number of resource bases or production plots (value: 1–11) and the number of types of resource bases (value: 1–6) accessible to and managed by the household are highly significant with respect to mean species count. Correlation

Table 5.2 *Summary table of significant factors affecting mean species count at the household level*

Variable	*F*-value	Correlation Coefficient	Probability
Age	0.11	-0.0046	0.90
Sex/gender	0.0.	-0.0038	0.97
Marital status	0.44.	-0.0861	0.64
Household size = *C*	0.03.	0.0006	0.973
Number of producers = *P*	0.26	0.0115	0.905
C/P ratio	0.78	0.1688	0.722
Number of resource bases	19.29[a]	0.6475	0.001
Number of types of resource bases	5.60[a]	0.1106	0.005
Vegetable sufficiency	0.66.	0.0144	0.520
Extra income	2.13	0.1253	0.101
With swidden	3.93[a]	0.0804	0.023
With home garden	0.12	0.0013	0.735
With livestock	1.13	0.0368	0.340
Area of landholdings	1.54	0.0331	0.220

Note: Unless noted, *F*-values are not significant.
[a]Significant at 1 percent level.

analysis shows that there is no relation between species counts on the one hand and household demographic characteristics, household size, or sources of extra income on the other. Nor is there a correlation between species counts and total size or area of all landholdings of the farmer (value: 0.25 to 14 ha), although from field observations it appears that farmers who "own" more land are more likely to leave one or more parcels on fallow as *natagwan*. Abandoned or neglected land can succeed to secondary forest growth or a woodlot, which may result in an increase in the number of woody species (in old fallows) or graminaceous species (in new fallows).

Cultivation of swiddens is negatively correlated with the mean species count of the household and is significant at a 1 percent level. Although swidden as a resource base exhibits very high taxonomic diversity, analysis at the household level shows that noncultivation of swiddens is associated with a higher mean species count for the household. Interviews suggest that a farmer who does not have a swidden plot cultivates more food crops on other kinds of land by way of compensation. Similarly, although home gardens as a resource base exhibit very high taxonomic diversity, statistical analysis at the household level shows no significant correlation between the presence of a home garden and the mean species count of the household. There is no significant correlation between livestock and mean species count at the household level, although interviews with household respondents indicate increased cultivation, when livestock are present, of feed and fodder crops such as sweet potatoes and other root crops and a greater tolerance for some graminoid species that are otherwise viewed as weeds but acquire value as fodder for livestock owners.

Analysis of Biodiversity at the Resource Base Level

A total of 323 plots of equal or similar size ($A \geq 0.5$ ha) and varying land use types belonging to ninety-three household respondents were surveyed for species counts. The present analysis is done by type of resource base, as each type differs in terms of physiography, land use, resource management objectives and strategies, and focus species. Table 5.3 shows the general trend in the mean species count of the different resource bases.

Table 5.3 *Mean species count for different resource bases*

Resource Base	N	Mean Species Count
Woodlot	3	23.30 (1)
Coffee plantation	44	9.30 (6)
Swidden	88	17.15 (3)
Rice terrace	81	11.28 (5)
Bean garden	42	14.14 (4)
Home garden	65	21.77 (2)

Note: Numbers in parentheses indicate rank from highest to lowest in the mean species count.

BIODIVERSITY INDICES. Data are presented in the following tables and figures: table 5.4 summarizes these biodiversity indices for ease of comparison among the different resource bases; table 5.5 presents indices of similarity and dissimilarity between the different resource bases; figures 5.3, 5.4, and 5.5 present vegetation maps and profiles of three resource bases, namely, woodlot, swidden, and home garden. These figures include information on plant associations, plant density, canopy structure, and tree architecture as well as the horizontal and vertical spatial distribution of species within each site.

Field observations and interviews at the study site revealed a composite swidden system that involves multiple land uses and optimum resource utilization within the landscape, including rice and vegetable production for home consumption and vegetable and tree crop production for market sales.

The communal resource bases—the forest *(ala)* and grassland *(kongo)*—were not included in the analysis of biodiversity at the household and farm/plot levels. Baseline information (i.e., species listing) on the forest and grasslands were, however, obtained at the landscape level during transect surveys.

THE WOODLOT. Landholdings visited during the study were identified and classified by the farmer respondents themselves. Of the total 323 plots belonging to the ninety-three household samples, only

Table 5.4 *Biodiversity indices for different resource bases*

Resource Base	Number of Species (*S*)	Total Number of Individuals (*N*)	Species Richness (*R*)	Shannon-Weiner Index (*H'*)	Evenness Index (*e*)
Woodlot	22 (3)	234 (6)	1.44 (2)	3.33 (2)	2.48 (2)
Coffee	21 (4)	566 (4)	0.88 (3)	2.02 (6)	1.53 (6)
Swidden	32 (2)	1,741 (1)	0.77 (4)	2.49 (4)	1.63 (5)
Rice or pond field	9 (6)	832 (3)	0.31(6)	2.03 (5)	2.14 (4)
Bean garden	18 (5)	1318 (2)	0.50 (5)	2.97 (3)	2.37 (3)
Home garden	57 (1)	523 (5)	2.49 (1)	4.64 (1)	2.64 (1)

Note: Numbers in parentheses indicate rank from highest to lowest value.

Table 5.5 *Similarity (IS) and dissimilarity (DS) indices among the different resource bases*

IS DS	(1) Woodlot	(2) Coffee	(3) Swidden	(4) Terrace	(5) Garden	(6) Home garden
(1) Woodlot	—	27.91	3.70	3.20	0	12.5
(2) Coffee	72.09	—	7.55	6.67	25.64	10.13
(3) Swidden	96.30	92.45	—	7.30	12.00	11.11
(4) Terrace	96.80	93.33	92.70	—	11.11	1.49
(5) Garden	100.0	74.36	88.00	88.89	—	0
(6) Home garden	87.50	89.87	88.89	98.51	100.0	—

three parcels were categorized as woodlots. (This sample is not sufficient to run statistical analyses of any significant differences with other resource bases.)

The small number of woodlots in the study area is the result of a recent conversion of many woodlots into market-oriented coffee and/or fruit tree plantations in response to market conditions. Additionally, maintenance of a woodlot by the farmer is a function of the total area of his landholdings. Farmers with more land are

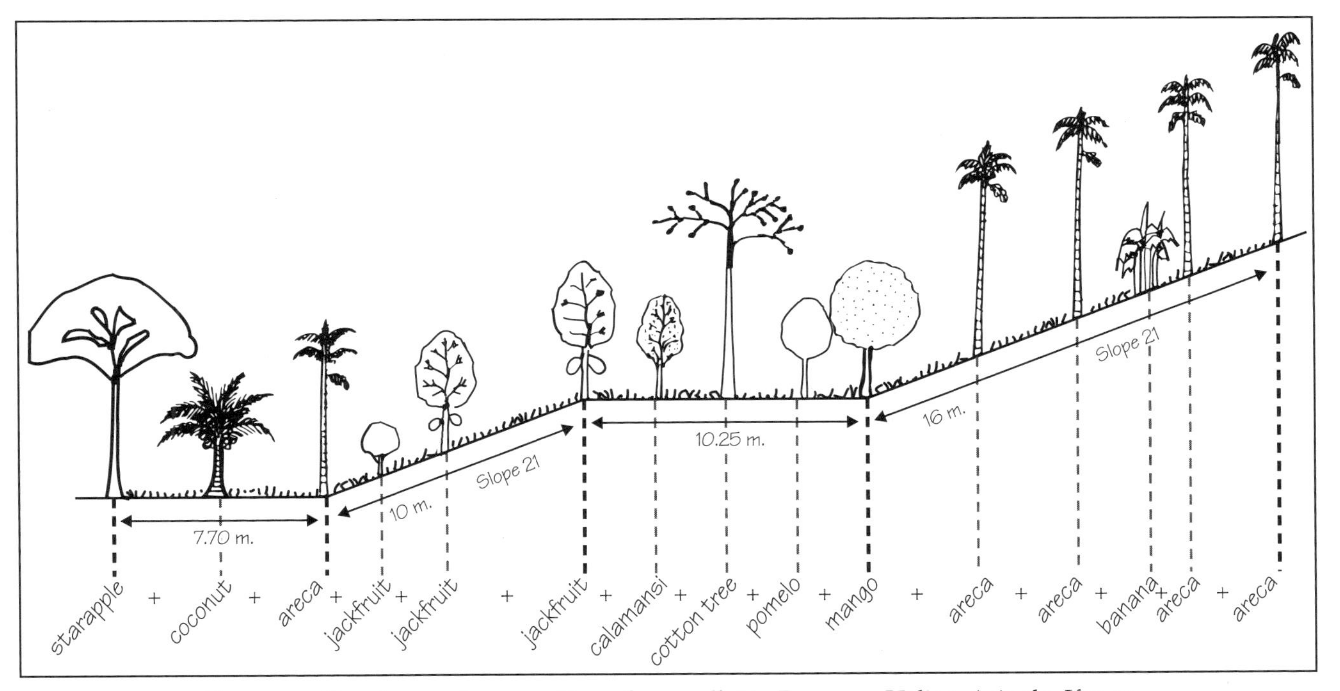

Figure 5.3 Structure and composition of a woodlot in Barangay Haliap, Asipolo, Ifugao

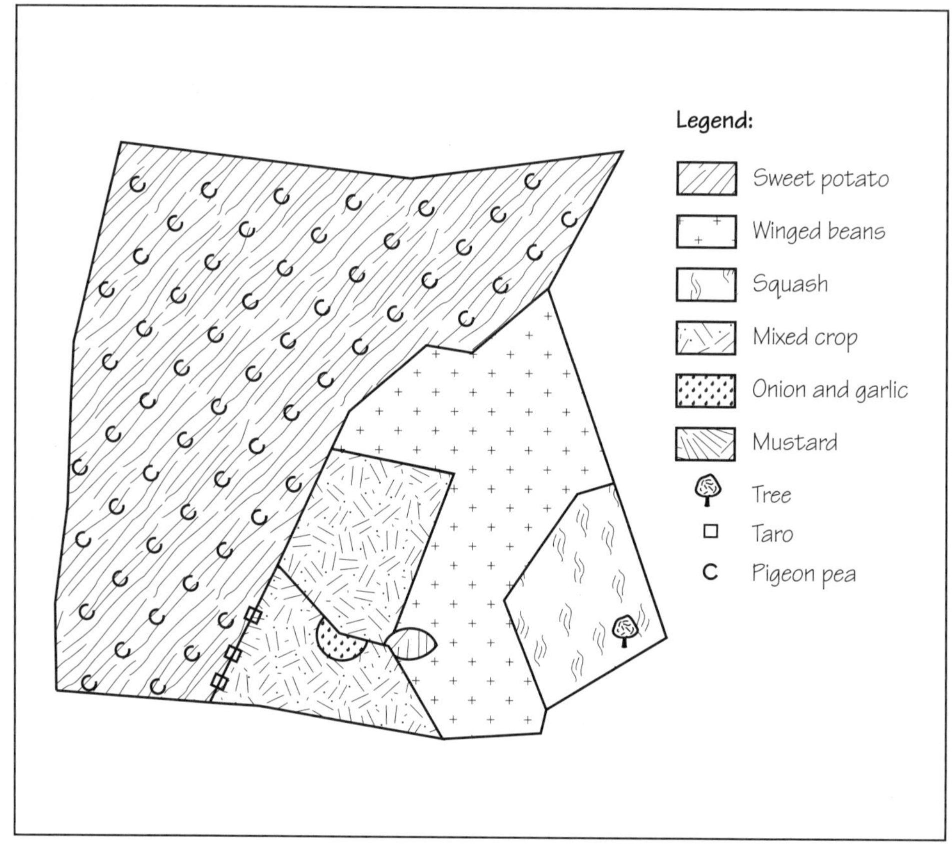

Figure 5.4 Vegetation in a *habal* in Sitio Li-cod, Barangay Haliap, Asipolo, Ifugao

more likely to leave a parcel as a woodlot until the need arises for other land uses or until a more beneficial alternative is perceived.

Potentially, woodlots exhibit relatively high species richness (both in types and frequency of species). Much of their biodiversity comes from the presence of originally wild perennials (from the original forest vegetation on the site) later enhanced by the planting of other tree species according to the household's preferences or needs. Tree species comprise 37 percent of the total number of species found in the woodlots and account for 80 percent of all the individual

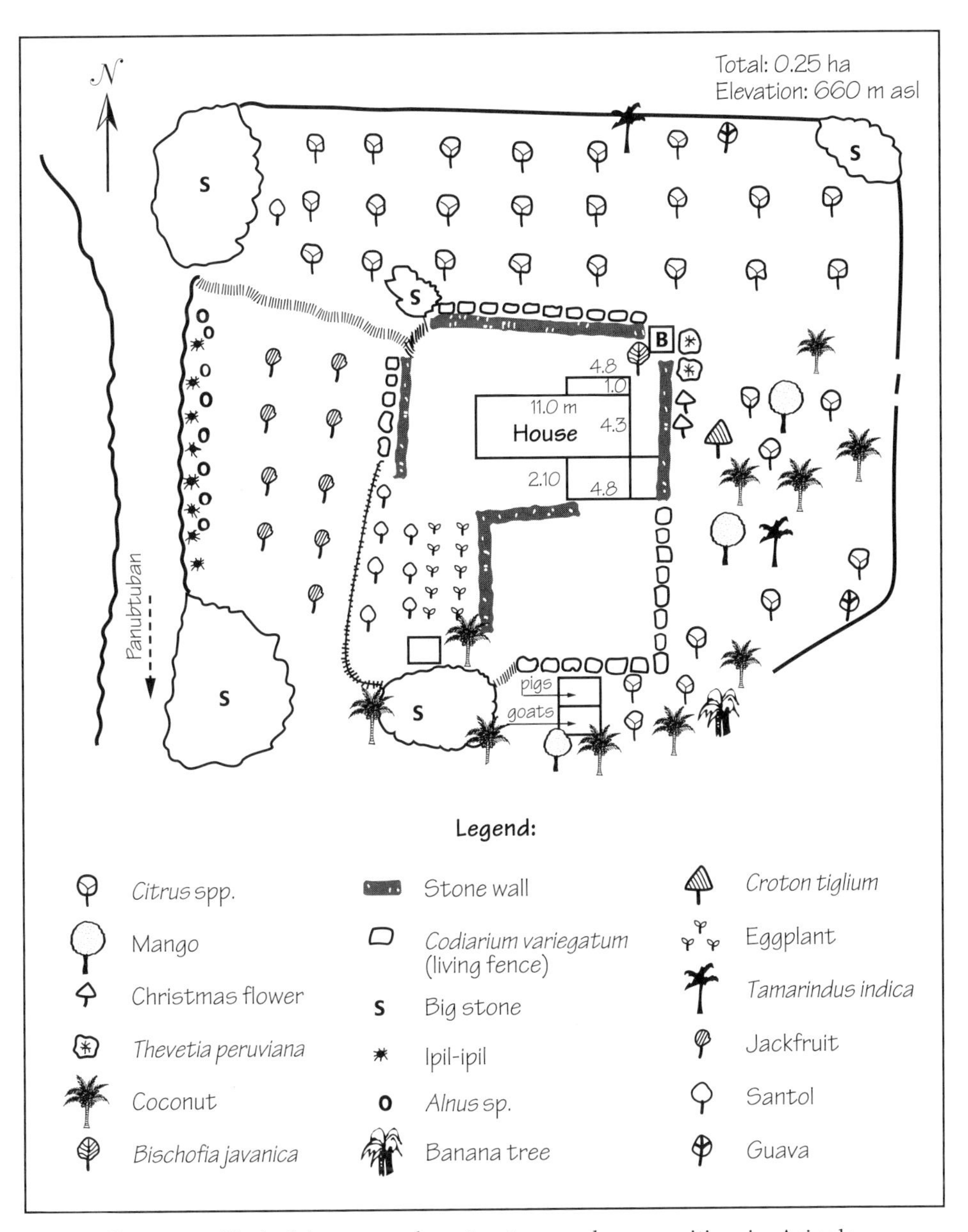

Figure 5.5 Typical home garden structure and composition in Asipolo, Ifugao

plants therein. Grass, sedge, and fern species are scarce, however, due to the shading effect of the more or less closed canopy of the woodlot.

THE COFFEE PLANTATION. A relatively low Shannon-Weiner index ($H' = 2.02$) was calculated for the coffee plantation, indicating a low biodiversity level relative to the other resource bases. This is an expected result for a monocropping plantations.

THE SWIDDEN. The basic strategy of Ayangan swiddening is diversification of cultivated crops (both food and feed crops), the maintenance and conservation of noncultigens perceived to be useful in one way or another (e.g., food, medicine, fiber, or erosion control), and maximum use of all space available. The success of this strategy is reflected in the high species count and number of individuals per species.

Interview data and participant observation demonstrate that a farmer who wants to expand his or her resource bases will prepare available land initially for a swidden, even if he or she currently has a swidden. Farmers also may enlarge the boundaries of an existing swidden by clearing the periphery and planting it to more or new food crops. Field observations and interviews further indicate that if the average Ayangan farmer can have only one type of resource base, he or she will cultivate a swidden plot, as this can provide the basic food (and feed) requirements of the household plus the potential of surplus for cash sales. Thus, there are more swiddens (in terms of number of plots) in Haliap-Panubtuban than any other resource base, which may be an indication of the Ayangan farmers' choice of swidden as a priority land use.

Species content of the swidden reflects high crop and weed diversity. Tree, shrub, and herbaceous species constitute 75 percent of all types of species found in the swidden, although relative frequency of individuals of these species may be low due to the predominance of legumes and other creeping vines (e.g., *Ipomoea* spp.) used as cover crop for soil protection.

THE RICE TERRACE. Except for one household respondent, all rice harvested from individual terraced fields is for home consumption.

Baseline data indicate that rice terraces have the lowest species count and species richness index and have a Shannon-Weiner index comparably low to the coffee plantation. Biodiversity in the rice terraces is very low due to the nature of the land use itself. Crop management calls for decreasing interspecies competition by eliminating weed species to maximize the yield of the primary crop (rice). Of significance, however, is the observation that one or more varieties of rice are planted within the terraces, not as a mixed bag of seeds but in separate subplots. The choice of varieties also is defined by end use (e.g., rice staple, rice wine, or quality of taste) as well as agronomic factors (e.g., differential grain maturity, resistance to pests, or drought tolerance).

Ayangan farmers cultivate other crops and small tree species (e.g., maize, winged beans, taro, or papaya) on the bunds of terraces and in any other available space, including on compost mounds in the field, primarily for home consumption. And, although rice terraces are intensively managed to prevent rice-weed competition, self-sown or volunteer species such as *Nasturtium indicum, Portulaca oleracea,* and *Amaranthus* spp. are tolerated for economic reasons because they can be eaten. This accounts for the presence of trees, shrubs, herbs, and vines in this resource base. Furthermore, as no chemical pesticides are used, fish and snails abound and are collected for home consumption.

THE BEAN GARDEN. In terms of biodiversity indices, results indicate that although species counts are lower in bean gardens compared to swiddens, the Shannon-Weiner index in gardens is slightly higher than in swiddens due to the lower species-to-individual ratio found in bean gardens. Trees, shrubs, herbs, and graminoids contribute about two-thirds of the total number of individuals while the main crop contributes the other.

THE HOME GARDEN. One would expect every household to have a home garden, as it is the area immediately surrounding the house cultivated or maintained for the production of a variety of annuals and perennials. However, results of the study show that 28 percent of the respondents did not cultivate or maintain the surrounding area of their homesites, owing mostly to a lack of secure land tenure where their houses are built. Unlike swidden plots, which may be "borrowed" and cultivated without security of tenure, homeowners who do not own their homesites do not have the motivation to develop the home garden, especially with perennial plants. At the most, these homeowners plant flowers and spices such as chili peppers. Following the aforementioned definition of a home garden, the area surrounding these residences is not considered a home garden.

The home garden exhibits the greatest biodiversity, not only in terms of taxonomic diversity but in terms of the other standard measures of biodiversity. Its high biodiversity status may be due to the fact that the home garden has become a general "dumping" ground for everything and anything. Ornamentals are intentionally cultivated for their aesthetic value, some spices and vegetables (chili, tomato, eggplant, onion) are planted within the vicinity of the house for quick and easy access at mealtime, and trees are planted to provide not only fruits or fuelwood but shade in the garden. Trees and shrubs contribute about 68 percent of the total number of species and about 65 percent of all individuals found, herbs and other vines contribute about 45 percent of the total number of species or 52 percent of all individuals found, and graminoids contribute minimally.

In this case, the resource base with the least intensive cultural management (no weeding, no land preparation, no fertilizer application) but the most frequent human visitation seems to yield the highest biodiversity.

Factors Affecting Species Diversity in the Resource Bases

Table 5.6 presents a summary of the factors affecting mean species count at the resource base level: land tenure, land acquisition, and stoniness. Of these three factors, only stoniness was statistically significant. There is very high negative correlation between stoniness of site and mean species count; the stonier the site, the lower is the species count. Stoniness affects the workability of the site and the demand for labor. The farmer tends not to plant many crops on stony sites, if at all. Furthermore, crop suitability is highly dependent on site quality. It was observed that very stony plots are planted solely to *Ipomoea batatas* (with minimal tillage), not as a major crop (the tubers are not harvested) but as a soil erosion measure.

Table 5.6 *Summary table of factors affecting mean species count in the resource base level*

	F-values					
Variable	(1) Woodlot	(2) Coffee Plantation	(3) Swidden	(4) Rice terrace	(5) Bean garden	(6) Home garden
Tenure	—	—	0.0	0.12	1.78	1.13
Stoniness	2.08	2.49	8.99[a]	0.97	0.18	2.78
Acquisition	—	2.0	0.55	0.78	0.84	0.70

Note: Unless noted, *F*-values are not significant.
[a] = Significant at the 5 percent level.

With respect to the effect of tenure, there is no statistically significant difference in mean species count in the different resource bases. From field observations, key informant interviews, as well as other studies in Haliap-Panubtuban (Manuta 1993), farmers who own land tend to plant more perennials (e.g., coffee and pomelo) and plant swidden areas to a variety of food crops for home consumption.

Likewise, there is no statistically significant variation between the mean species count of the resource base and the manner of land acquisition. Acquisition may occur through inheritance, purchase, rental, mortgage, pioneering, borrowing, or gift. Although they are of no statistical significance, inherited woodlots exhibit the highest mean species count. This is probably due to the length of time it was a woodlot and the inherent high species diversity of this type of land use.

Plant Indicators of Biodiversity

Aside from the physical and socioeconomic variables described and correlated in the previous sections, another result of this study is the identification of specific plant species that can be used as surrogate indicators of biodiversity levels. Based on interviews and field observations, it was determined that certain activities involved in specific land uses (coupled with conditions such soil fertility status) may also provide good indications of the biodiversity levels at the farm or resource base level. The presence (though not the absence) of each identified plant species in a resource base was tested for statistical significance with respect to the mean species count as the dependent variable (as the measure of the level of biodiversity) of that particular resource base. It should therefore be pointed out that the identified species may be an indicator of the biodiversity level in one resource base but not in another. Furthermore, statistical correlation analysis only indicates significant differences in biodiversity levels; it does not indicate actual biodiversity levels (i.e., low vs. high levels of biodiversity). Five species were identified as significant indicators: *mom-ma* or betel nut palm *(Areca catechu)*, *oshe* or pigeon peas *(Cajanus cajan)*, *lagka* or sweet potato *(Ipomoea batatas)*, *gulon* or *alang-alang* grass *(Imperata cylindrica)*, and *lubfuan* or pomelo *(Citrus grandis)*.

"ARECA CATECHU" ("MOM-MA," BETEL NUT PALM). Betel nut chewing is a favorite activity among the Ifugao people and is as popular, if not more so, as smoking or chewing tobacco. The nut is important in rituals as an offering to the spirits, who are believed to enjoy chewing nut in the other world. The trunk of the areca palm is considered by the local people to be an important, hard, and long-lasting construction material; it does not rot even when wet and is known to have inherent insecticidal properties. It is the preferred material for flooring in outhouses. From field observations, it was noted that Areca palms are planted in woodlots, coffee plantations, swiddens, and especially in home gardens. It was gathered from interviews that farmers who have a strong perception of security of tenure in their landholdings plant Areca palms to indicate ownership and mark the boundaries of their property. The tree's longevity and characteristic tall and erect trunk make it highly visible from a distance and, as such, an excellent boundary marker. Furthermore, selected case studies indicate that farmers who own land tend to plant a greater variety and more trees on their property for future use, resulting in greater biodiversity in the production plots. Statistical analysis shows that the presence of Areca has a positive correlation with mean species counts in swiddens, rice terraces, bean gardens, and home gardens.

"CAJANUS CAJAN" ("OSHE," PIGEON PEAS). The presence of pigeon peas (depending on their growth stage) indicates that a swidden is in its second cropping stage. Statistical analysis shows positive correlation with mean species count in swiddens, bean gardens, and home gardens. It is during the second cropping stage that farmers diversify plant production; the first cropping stage is usually for upland rice production. It follows that the second cropping stage is more diverse than the first.

"IPOMOEA BATATAS" ("LAGKA," CAMOTE, SWEET POTATO). The presence of sweet potatoes in an expanse of cultivated land indicates that the land or plot is in its last stage of swidden cropping, which by then may have a great variety of food and fodder crops as well as volunteer species. Statistical analysis shows a positive correlation between the presence of *Ipomoea batatas* and mean species count in the swidden and the home garden. From key informant interviews and selected case studies, it appears that *Ipomoea* is planted if one has livestock, such as pigs, indicating the need for feed. Along with sweet potatoes, other crops such as cassava and taro are planted in the swidden (and the home garden) to provide a variety of fodder crops for the livestock.

"IMPERATA CYLINDRICA" ("GULON," "ALANG-ALANG" GRASS). The presence of *Imperata* indicates that the area has been extensively and repeatedly burned after intensive cultivation of food (swidden) and cash (garden) crops. Its presence has always been perceived as an indicator of decreasing soil fertility and a signal for permitting the plot to lie fallow. Statistical analysis indicates, however, that the presence of *Imperata* is positively correlated to mean species count in the bean garden. This could probably mean that the bean garden had been invaded by weed species (e.g., *Imperata* and others) following the lack of intensive weed management. This condition could make the plot high in species content than when *Imperata* is absent or weeded out. In swiddens, the invasion of *Imperata* signals the time to abandon the plot until such time when *Miscanthus* succeeds *Imperata*.

"CITRUS GRANDIS" ("LUBFUAN," POMELO). This tree, favored for the high market price of its fruit, is usually planted in privately owned and managed woodlots or home gardens. Its presence, like that of *Coffea* spp., indicates the tenure of the plot. Statistical analysis indicates a positive correlation with mean species count in the swidden and bean garden. This may further indicate new conversion from old fallow to either swidden or bean garden and would therefore still contain some perennial species, especially fruit-bearing types.

Results of the study therefore indicate the usefulness of certain plant species to mark high or low levels of biodiversity in the production plot or resource base by indicating other characteristics of the system, such as tenure security, stage of fallow, or cropping season, that are also correlated with biodiversity levels of the system (i.e., sociobiological or composite indicators). Therefore, these plants can be used as rapid indicators of biodiversity in the anthropogenic landscape.

Biodiversity as an Indicator of Sustainability: A Functional Analysis

Sustainability is a measure of whether the productive potential of a certain natural system will continue under a particular management regime. Indicators of sustainability have been suggested by a number of workers using energy use, biological wealth, policy, institutions, society, and culture as attributes (Carpenter 1990). The present study also aims to present biodiversity (and its conservation) as an indicator of sustainability of the productive potential of an indigenous agro- and natural ecosystem with the Ayangan of Haliap-Panubtuban as a case study, focusing on the functions of biological diversity (i.e., the ecological, economic, and sociocultural functions of plants).

Economic functions include the plant's role as food, feed, drink, medicine, fiber, timber, and other direct consumptive values. Sociocultural functions include ritual and symbolic uses, religious significance, sports and recreation, and ornamental and other aesthetic uses. Ecological functions are those that contribute to environmental integrity and ecological processes such as soil and water conservation, nutrient cycling, succession and biotic stability, seed dispersal, germination, and pollination, among others. The type of biodiversity (e.g., tree, shrub, herb, or graminoid) retained or managed becomes an indicator of sustainability of the production system at the landscape level.

The dependence of local residents on diverse biological resources is an important factor influencing biodiversity conservation. To the Ayangan of Haliap-Panubtuban, biodiversity and its conservation is highly use defined. The ecological functions of a species may or may not be perceived by the farmer, but its perceived economic and/or sociocultural value may enhance its maintenance and conservation. Perceptions of economic and cultural values, however, may be influenced by factors exogenous to the community and landscape, such as market and development interventions, which may result in a general change in personal values, attitudes, and general use definitions.

Regardless of whether changes in perception have already occurred, a functional analysis of biodiversity is proposed using some examples from the results of the ethnobotanical survey (see table 5.7). The analysis is intended to assess conditions and trends in maintaining and/or conserving certain species of plants essential to sustaining the Ayangan-Ifugao production system in the Haliap-Panubtuban landscape.

Implications of the Analysis of Biodiversity at the Landscape Level

MULTIPLE USE STRATEGY: SPATIAL AND TEMPORAL INTERRELATIONSHIPS IN THE LANDSCAPE. The traditional Ayangan production system is swidden based, in the sense that all productive activities of the household (i.e., farm, off-farm, nonfarm) are synchronized with the swidden calendar. It also follows a multiple use strategy in managing natural resources, tapping several different types of resource bases in the Haliap-Panubtuban landscape at the same time to produce an uninterrupted and mixed flow of products to meet the household's cash and subsistence needs. This strategy is based on diversification of production through the creation of "patches" in the landscape, which, when analyzed individually, may appear specialized and monocultural and generally low in biological diversity. Coffee plantations, rice terraces, and bean gardens are each specialized land uses (monocultures, focusing on one crop) but their integration with swiddens, woodlots, and homegardens into a

Table 5.7 *Functional analysis of biodiversity in the species level (an example)*

Latin Name, Local Name, Family name	Plant Habit	Biodiversity Function			Remarks
		Economic	Sociocultural	Ecological	
Bischofia javanica "tuao" Euphorbiaceae	Tree	Source of good-quality timber for building construction, furniture making, and woodcraft	Holy tree	Perennial; deep rooted; serves as host to several epiphytic species; microhabitats	Common and numerous; found not only in woodlots and coffee plantations but in swiddens and homegardens
Macaranga bicolor "anablon" Euphorbiaceae	Tree	Fiber used for bundling material during rice harvest	Distinct red color of fiber used to make traditional/ceremonial clothing	Perennial; grows in a variety of habitats	Although occurring in the wild, sometimes cultivated in woodlots and fallow plots
Ipomoea batatas "lagka-lagka" Convolvulaceae	Tuber crop; creeping vine	Tuber used as food and feed; leaves edible as green vegetable	Staple food of the "*Camote* Eaters"	Used as cover crop during fallow of swidden plots for soil erosion control	Several varieties are maintained/planted on the same farm depending on farmer's use; high genetic diversity
Areca catechu "mom-ma" Palmae	Tall, erect palm	Masticatory; nuts for personal use and market sales; trunks make durable construction material	Betel nut chewing is not only an important leisure activity, but highly significant as ritual offerings to make "peace with the spirits"	Perennial	An *Areca* palm indicates human occupancy or tenure of that particular site, which can further indicate some sort of management of the site

(continued)

Table 5.7 *(continued)*

Latin Name, Local Name, Family name	Plant Habit	Biodiversity Function			Remarks
		Economic	Sociocultural	Ecological	
Miscanthus sp. "runo, filao" Graminae	Tall cane/grass	Long canes used as trellises for beans; leaves fed to cattle	Believed by farmers to indicate restored soil fertility in fallow plots	Can successfully take over *Imperata cylindrica*, a noxious weed	Commercialization of *runo* canes indicates diminishing grass cover of grasslands in the landscape
Ficus benjamina "balete" Moraceae	Tall tree with adventitious roots	Source of fuelwood	Holy tree believed to be inhabited by evil spirits	Perennial and deep rooted; fruits eaten by birds and bats	Common but not logged; host to a variety of vines and epiphytes
Cyathea contaminans "atibfang" Cyatheaceae	Tree fern	Trunk useful as construction material	Trunk carved to make rice gods (icons)	Pioneer species after a burn; resistant to termite attack and dry rot; used as a repellent against stink bug rice pest	Common and numerous in the area

larger unit of production (i.e., at the landscape level) provides a wider variety of resources and biological diversity for the household.

The Ayangan of Haliap-Panubtuban are a community in transition with a developing market orientation. Land use management depends not only on food security but on meeting other cash and material needs, primarily children's education and health care. It tends to rely on resource bases (e.g., swidden, coffee plantations, and bean gardens) that can provide the highest returns on invested labor and capital.

Diversification of resource bases and the resultant increase in biodiversity at the landscape level are effective buffers against market fluctuations and environmental changes and hazards (e.g., typhoons, earthquakes, and pest outbreaks) for the Ayangan farmer. Swidden produce provides the household with security against pond field damage, crop failure, and the steady decline in coffee prices and the fluctuating market price of beans. The cultivation of fruit trees also provides additional products for the market. Note further that land use conversion from rice terraces to bean gardens or diversification into another resource base was initially a response of the Ayangan farmer to environmental stress brought about by the prolonged drought in the region, which caused crop failures in the rice terraces.

Management strategies for each resource base involve maximizing the use of available land. The bunds of rice terraces and the margins or peripheries of bean gardens are not left bare; they are either planted with other domesticates or allowed to grow edible weed species.[2]

The steep and stony slopes of swiddens are planted with sweet potatoes. This practice may appear too intensive, but the choice of crops makes a difference in the strategy. According to the local people, sweet potatoes, for instance, require good drainage for maximum tuber growth. At the same time, their extensive root system and creeping vine and broad leaves provide good soil cover, protecting against surface erosion. Planting other domesticates on the bunds of terraces serves a similar soil conservation function and at the same time provides food for the household. This multiple use strategy therefore provides the management unit with its basic subsistence requirements while conserving the soil.

INDICATORS OF SUSTAINABILITY IN THE HALIAP-PANUBTUBAN LANDSCAPE. Historical data and results of the current study indicate that the present production system of the Ayangan of Haliap-Panubtuban has been sustainable through the years. The traditional resource management and resource use strategies and other activities in the two subsystems of the agroecosystem (i.e., the biological and human subsystems), as reflected in the multiplicity of production plots, have maintained high levels of species diversity at both the farm and landscape levels.

Bio-socioeconomic indicators show similar sustainability in the present production system. First, soil-water conservation measures such as cover cropping, crop rotation, alley cropping for ground cover, stone walls, and terracing are widely practiced by a majority of the farmers in most production plots, minimizing soil loss, especially on sloping lands. Plant indicators found in the present study indicate, for instance, the planting of certain species (e.g., *Ipomoea batatas)* in sloping, stony, or production plots to be left fallow to minimize soil erosion as well as enhancing fallow by planting perennials (e.g., fruit and timber species). Not to be forgotten are the rice terraces, which brought the Ifugao to fame. Second, the length of the fallow period after swidden cropping is between five and fifteen years, allowing natural succession to proceed and soil fertility to increase. Third, external inputs of chemical fertilizers and pesticides are zero to minimum, with most farmers using only natural or composted materials to fertilize their production plots (except in bean gardens). The abundance of edible snails in the pond fields and waterways is also a useful indicator of the health of these resource bases. Fourth, labor distribution and decision-making processes involve family participation of household members of all ages and genders in various farming activities. Fifth, food security (i.e., food availability and accessibility) for the household as a result of the system's yield or productivity includes a minimal surplus that can be sold to meet the minimum cash requirements of the household. Sixth, water quantity and quality are sufficient, although the site has experienced water stress in the more recent years, for which farmers blame the severe drought and earthquake of 1991. Traditional farmers reverted to previous land uses (i.e., rice terraces and swidden) as soon as water became available in the rainy season. Barring the effects of the drought on the water supply, trends in water quantity for agricultural use have not changed drastically over the years as manifested by the high number of production plots that require water: swiddens, which are rainfed; and rice terraces, which are irrigated. Rivers continue to provide irrigation water and no respondent recalled them ever drying up. Children and adults can still be observed bathing, swimming, or fishing in deeper portions of the rivers, indicating good water quality. The availability of such edible water plants as *Nasturtium officinale* (watercress) on smaller waterways and edible snails in pond fields also indicates good water quality at the site. The water supply for domestic use at the site has actually improved in the last four years as a result of spring protection and impoundment structures built by community members with the assistance of various development organizations (e.g., the Central Cordillera Agricultural Program). Several *sitios* in Haliap and Panubtuban have piped water in residential homes. Boiling water for drinking is unnecessary, and reports of water-borne diseases are few. The presence of bananas (*Musa* spp.), a plant that has a high water requirement, in most fallow plots indicates that the area has ample water for agricultural activities.

Although recent trends indicate that there are high biodiversity and sustainability levels in the present production system, certain factors exogenous to the agroecosystem may bring about a decline in biodiversity and consequently in the sustainability of the production system. These factors include market influences such as the high cash value of certain crops, which may bring about extensive land use conversion to monoculture of crops such as beans.

The loss of biodiversity at the household level as a result, for instance, of total conversion of all household production plots into bean gardens may leave the farmer with no fallback crop, even for household consumption, in case of a market crash or crop failure or due to pests, diseases, or natural hazards such as severe floods, droughts, and typhoons. As pointed out earlier, commercial bean gardening is risky due to the unpredictable fluctuations of the market. A fall in the market price of beans can leave a farmer deep in debt. Livestock production (e.g., of pigs, chickens, goats, and ducks) would also be affected. With no fodder crops available, meat would also become scarce. The effect of the intensive use of chemical fertilizers and pesticides on the soil's chemical-physical condition and microbiota would make it difficult to revert to the original land use and cropping systems. Soil loss would also be massive, as the total conversion of production plots to bean gardens involves draining rice pond fields and neglecting the rock walls of the terraces. Intensive weeding of bean gardens would leave only minimal ground cover, resulting in surface and gully erosion. The fallow period would not be sufficient for the soil to rest and recover its natural fertility, as beans can be cropped three times a year. Labor distribution and decision making would be lopsided, leaning toward the male members of the household, as intensive agricultural labor—land preparation, pest management (chemical spraying), and hauling produce to market—is perceived as the male's domain.

Other external variables, such as improvements in the road infrastructure, may also influence land use conversion to bean gardening, as ease of transport of produce or market access would improve. This second factor may already be operating. Most bean gardens are located in Barangay Haliap, and Panubtuban farmers have expressed interest in bean gardening when the road between Haliap and Panubtuban is improved.

Conclusion

Biodiversity has become a buzzword, and biodiversity studies were proposed and implemented in most types of ecosystems in most parts of the world even before the 1992 Rio Summit. Most biodiversity studies, however, dwell on the conditions that bring about loss of biodiversity or the conservation of a rare, threatened, or endangered species. Biodiversity studies, especially those whose main objective is conservation, have focused their attention mainly on the nature of what is being conserved and have seldom considered the viewpoint of the people dependent on such resources for their livelihoods or the conditions associated with local people's management strategies that may result in successful biodiversity conservation.

Biodiversity studies commonly use such standard indices as species richness (*R*) and Shannon-Weiner indices (*H′*), which place numerical values on biodiversity levels. Although these standard measures give information on type and distribution of species present in a particular sampling area, they seldom explain the causes or factors affecting numerical values such as why these species are present at all, why in such quantity, why in association with other species, and why in such an area. The importance of the rapid appraisal methodology developed here lies in its ability not simply to compute biodiversity indices but to identify surrogate parameters that may indicate the overall conditions of biodiversity. This approach to studying biodiversity facilitates more rapid research and can answer specific qualitative questions such as those mentioned above. Furthermore, this approach enables policymakers to gather initial, relevant information on biodiversity without having to invest in time-consuming and costly inventories of species.

Studies should not view species diversity only in terms of biology and/or taxonomy. The analysis of biodiversity indices at different hierarchical levels (species, farm/household, resource base, and landscape) should consider the economic, sociocultural, and, importantly, ecological values, functions, and contributions of biodiversity across the landscape. Biological diversity often coexists with human cultural diversity, as the latter is built and largely dependent on what the former has to offer, both materially and philosophically. Studies on biodiversity should therefore consider information on the human component, including analysis of such household characteristics as socio-demography and tenurial status, including resource management objectives such as monocropping, and strategies such as crop associations, soil and water conservation measures, and livestock management. Likewise, analysis of biodiversity at the farm or resource base level can highlight the contributions of biodiversity to the structural attributes (e.g., site quality or degree of patchiness) and functional attributes (e.g., impact on genetic erosion, impact on species endemism, or endangerment and/or extinction) of the landscape.

In the Haliap-Panubtuban landscape, the production system of the Ayangan can be sustainable if biodiversity is conserved in the crop production management strategies (i.e., indigenous and integral traditional agroforestry and inter- or multiple-cropping systems) that focus on perennials, shrubs, and herbaceous species, which can reconcile the economic, social, and ecological requirements of both the upland landscape and the upland farm family or household. This study provides data to support the notion that when patches of agricultural and agroforestry systems are viewed not as islands of human disturbance in a "natural" ecosystem but as components of diversity at the landscape level, the potential of a complex anthropogenic system to contribute to overall ecosystem health and species diversity can be assessed (See Puri, Sajise et al. and Gunawan et al., this volume). This study also demonstrates the need to apply the rapid biodiversity appraisal methodology at other sites to further improve and refine their applicability.

The data from this study are too preliminary to predict how long the complex agroecosystem of the Haliap-Panubtuban region will continue to be successful in supporting a diverse ecosystem at the landscape level. Particularly, as increasing market pressures influence the growth of monocultures and state-level policies promote more intensive cultivation of market-oriented species there is potential for this agroecosystem to become unsustainable, as demonstrated by Gunawan et al. (this volume). This issue should be a central concern of policymakers as they consider how governmental programs can support or alter the current agroecosystem. This study, therefore, is an essential first step in understanding the role of local peoples in modifying the landscape. It should be very useful to policymakers as they consider the broad impacts of increased dependence on market-oriented crops and the loss of the traditional composite agroecosystem.

Notes

1 This land-use classification was similarly described by Conklin (1980) in his ethnographic atlas of the Ifugao.

2 A weed is a plant that is not valued where it is grown. It is an unwanted species mainly because it competes with the cultivated species.

6

A Landscape Fragmentation Index for Studying Biodiversity Conservation Using GIS

DANTE K. VERGARA

THIS CHAPTER INVESTIGATES the effects of landscape patchiness on levels of biodiversity and endemism in a forested reserve that contains farms and settlements within its boundaries. Throughout, the terms *fragmentation* and *patchiness* refer to the same thing, namely, clearings in a forested setting, human influenced or not. Tree falls and inhospitable soils or slopes represent nonanthropogenic sources of patchiness in the forest (Forman and Godron 1986). Essentially, therefore, this chapter deals with nonforest patches or fragments in a forested landscape.

The results of this study are particularly relevant to habitat fragmentation resulting from agricultural activity on the borders and within state forest reserves. Despite national laws that make human use of state forests in the Philippines illegal, centuries of agricultural activities in the region will not be easily halted. Therefore, it is prudent to consider the best ways to plan and manage forest reserves so as to maximize biodiversity conservation while including the activities of resident settlers. The findings of this study suggest that if farmers' clearing can be distributed in sparse, discontiguous, and regularly shaped patches, the negative effects of habitat fragmentation can be ameliorated.

The chapter begins by reviewing the only existing fragmentation index and explaining its inadequacy for this investigation. It then proceeds to the development of a new fragmentation index and the methods involved in deriving it from maps and remotely sensed data using geographic information systems (GIS) and image-processing systems. Finally, the results are presented of regression analyses carried out to examine the relationship between this index, and the biodiversity indices and endemicity ratios generated in other studies conducted in the Mount Makiling Forest Reserve in the Philippines (see Sajise and Ticsay-Ruscoe 1997; Gruezo 1997; Gonzalez and Dans 1997; Gruezo and Gonzalez 1997; Francisco 1997).

Landscape Fragmentation Index

Definition of Terms

Before turning to the discussion of the indices themselves, a few words on the definition of terms used in geographic information systems and image processing are in order. Spatial data, or information regarding space, can be represented in two ways. The first, called raster representation, or simply raster images, consists of the digital representation of spatial data. Raster images are produced in much the same way as a fax or photocopying machine prints out a graphic page: one row at a time, from left to right, top to bottom. The image is made up of tiny white, black, or colored dots or cells, organized in rows of fixed width, such as you would find in a framed cross-stitched picture. Each dot, colored or not, is referred to as a "pixel," short for "picture element." Each color is referred to as a "class," and different GIS and image-processing systems have different capabilities as to the maximum number of classes they can handle. Areas of contiguous pixels of the same class or color are called "polygons." Polygons can have holes in them of a different class or color, and these holes are called "islands." The whole collection represents the image, any subset of which is called a "window" of that image.

The second means of displaying spatial data is by vector representation, which is often referred to as vector images. This is the more intuitive means of representation and is akin to line drawings in which one picks up a pencil and begins to draw on a piece of paper. Each stroke of the pencil, called a "line," is made up of long or short straight edges called "arcs." Lines that return to their

d, and quadrat coefficient *Q* in the formula for the fragmentation index *f*, the fragmentation index can only assume values in the range from and including zero but less than 1.

Methodology and Analysis

Vegetative Characterization of the Makiling Forest Reserve and Its Environs

The Mount Makiling Forest Reserve is dominated by wet tropical forest. On the fringes of the forest and on the lower slopes of the mountain there are grasslands and agroforestry areas made up of both multistoried structures of mixed forest and fruit trees, and simpler structures consisting of mixed plantations of coconuts and/or fruit trees sometimes with understory agriculture. The plantations are usually cleared of weeds during the dry months of March and April in preparation for the planting season at the onset of the monsoon rains in June. The grasslands are dominated by *Imperata cylindrica* and *Sacharrum* spp. interspersed with some agricultural fields or fruit tree plantations. On the mildest slopes, there are agricultural fields planted to sugarcane or pineapples, while on completely level areas there are usually rice paddies and settlements.

Data Acquisition and Preparation

This analysis is based on the use of a Landsat 5™ quarter image taken at 1:38 P.M. on April 2, 1993, which was rectified for image distortion at the National Mapping and Resource Information Authority of the Philippines. A 400 × 400 pixel subset of the image was used, with each pixel covering approximately 30 × 30 m on the ground. To improve contrast, each of the seven bands was "stretched" by dropping 2.5 percent of the least significant pixels at the extremes of the spectral responses before any compositing was conducted. Topographic maps (1:50,000 in scale) of the area were obtained and digitized to generate the digital elevation model and the drainage network or the system of streams and waterways. They were also used to geo-reference or superimpose a known geographical reference system on the image using the latitude-longitude reference system in decimal degree units. Geo-referencing was performed by obtaining the coordinates of the highest peak from the contour map and affixing this to a single pixel patch in the image corresponding to the only bare spot at the exact peak of the mountain. The accuracy of this fix was tested by taking two recognizable points in the image corresponding to a road crossing and a distinct road curve that were near the edges of the image but diametrically opposite each other. The estimates were in error by at most 2 pixels, which amounts to just one-half of 1 percent error for an image of 400 × 400 pixels. The resulting digital image was used for all other analyses. The land cover/land use map was derived from this image after it was fully classified, and this was reclassed into forested and nonforested areas for the fragmentation analysis.

Ground-truthing was conducted by visiting the sampling sites and taking geographic coordinate fixes, wherever possible, using a Magellan Fieldpro V geographical positioning system (GPS) unit. It was not possible to get a GPS "fix" or reading in the closed canopy forest; this was only possible in areas with breaks in the canopy cover. The faunal transect lines (which ran through the floral sampling points) were estimated using the GPS fixes wherever possible; otherwise, they were estimated using the digital elevation model and drainage system map. The survey data for the Mount Makiling Forest Reserve were obtained from the Institute of Forest Conservation, University of the Philippines at Los Baños, and used in the computation of the boundaries of the reserve. The classification of the land cover/land use map was also greatly aided by the author's familiarity with the mountain.

Image Processing and Interpretation

The IDRISI geographic analysis system was used for the GIS analysis, as well as for image processing (Eastman 1993, 1996).[2] A principal component analysis over all seven bands, which results in a composite image that explains the most variation in the original image and is especially well suited for vegetation studies, was used to select the initial sampling sites before full classification of the image could be performed. A normalized difference vegetation index map, a technique used to describe the relative amount of green biomass in the image, using the near infrared and red bands, was also generated.

For the land cover/land use map, it was essential to discriminate between forest and coconut/fruit tree plantations and between grasslands and adjacent agricultural areas. But neither the principal component analysis, the normalized difference vegetation index, nor any of the usual techniques for discrimination of vegetation or land use types afforded the needed differentiation (Lachowski, n.d.). The band 5, band 4, and band 3 composite in the blue, green, and red channels overestimated the forest, as did the principal component analysis and the normalized difference vegetation index, while the band 3, band 5, and band 7 composite overestimated the patches. But our first hand knowledge of the canopy structure and undergrowth characteristics and the coincidence between the time the image was taken and the annual weeding season led us to attempt to use the thermal band to start a general classification. By combining near and midinfrared bands 4 and 5 with thermal band 6, the resulting composite (456) successfully differentiated the general areas for forest, mixed agroforestry and grasslands, and mixed coconut/fruit tree plantations. Although the poorer resolution of the thermal band (120 × 120 m) strongly influenced the composite and initially posed a problem by overemphasizing the shapes and sizes of the patches, this was overcome by overlaying selected supervised classes of the other composites with selected supervised classes of the 456 composite. The mossy forest, described in the literature at 900 m above sea level, was delineated using the digital elevation model. The classification scheme and process flow are summarized in figure 6.1, while figure 6.2 shows the classified land cover/land use map.

GIS Data Input and Analysis

To begin the fragmentation analysis, the land-cover/land-use map was reclassified to show the nonforest polygons on a forest background. Seven floral and five faunal sampling sites from various studies sites in the Mount Makiling Forest Reserve in the Philippines were used for this analysis (see Sajise and Ticsay-Ruscoe 1997; Gruezo 1997; Gonzalez and Dans 1997; Gruezo and Gonzalez 1997; Francisco 1997). The five faunal sampling sites corresponded exactly to five of the seven floral sampling sites. Of these five sites, four had transect lines for the faunal studies that passed through the centers

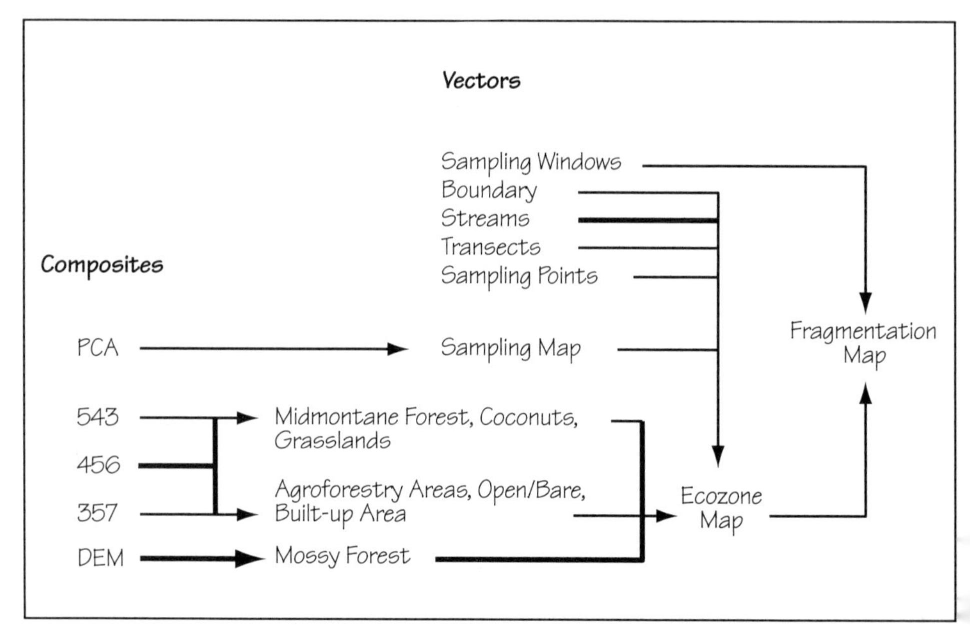

Figure 6.1 Classification scheme and process flow

of their corresponding floral sampling plots. Two of these transects were located in the midmontane forest, one in a grassland and the other in an agroforestry area. The fifth site, which was located in the mossy forest at the peak of the mountain, did not have enough space for a transect line. The last two of the seven floral sampling sites were located in another grassland and agroforestry area, respectively, and had no corresponding faunal data.

The midpoints of the four transect lines and the centroids of the sampling plots in the remaining three sampling sites were estimated and used as the centers for the sampling windows in the fragmentation analysis. The sampling windows vary from 33 × 33 to 34 × 34 pixels in size, with the majority within 33 × 34 pixels or approximately 1 sq km in size. (This variation is about ∀ 2.5 percent and thus tolerable.) The size of the sampling window was based on the knowledge that the faunal transects range from 1 to 2 km in length and run through the centers of the floral sampling plots. It was thus confidently felt that the data reported in these previous studies (see Sajise

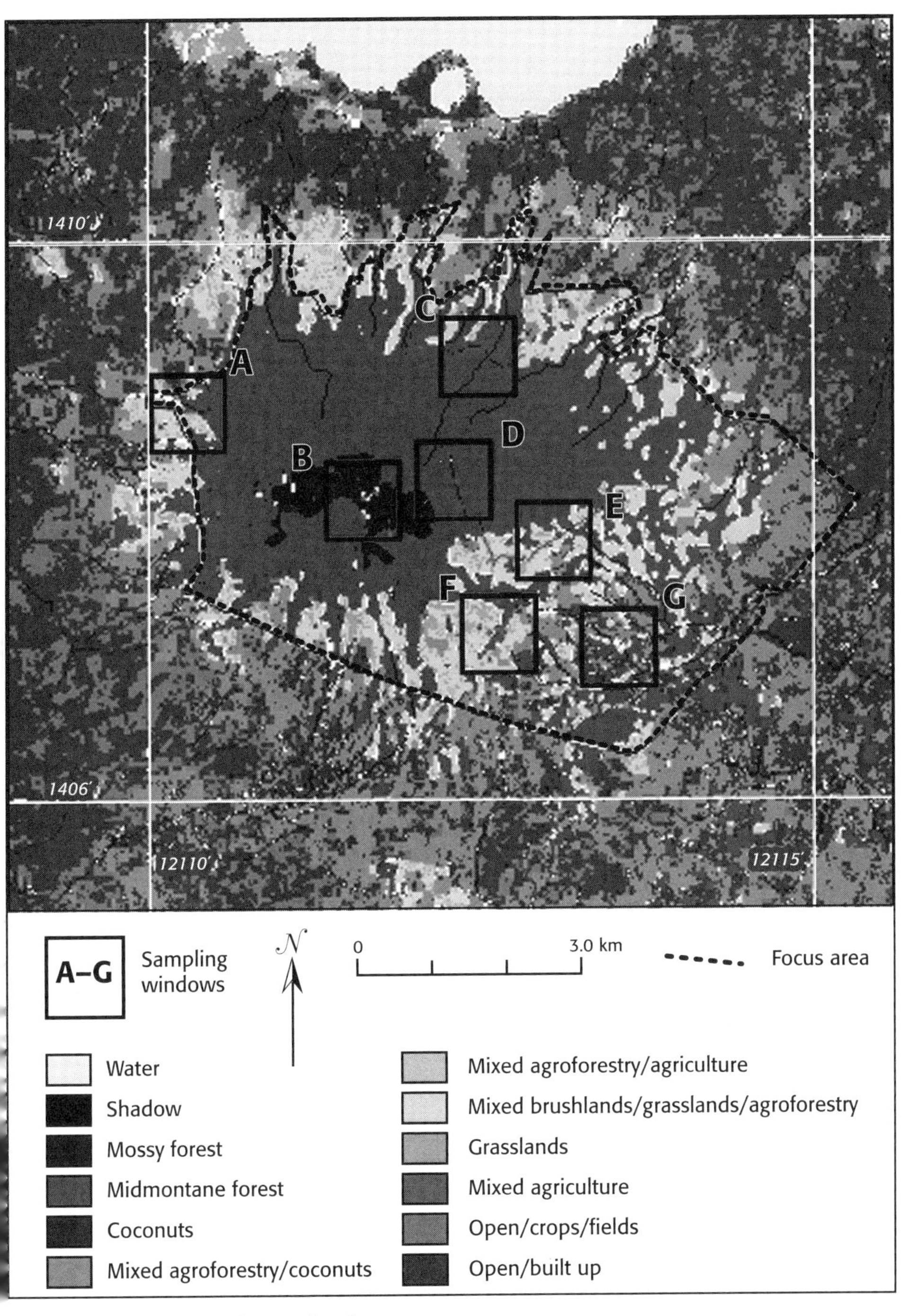

Figure 6.2 Land cover/land use map

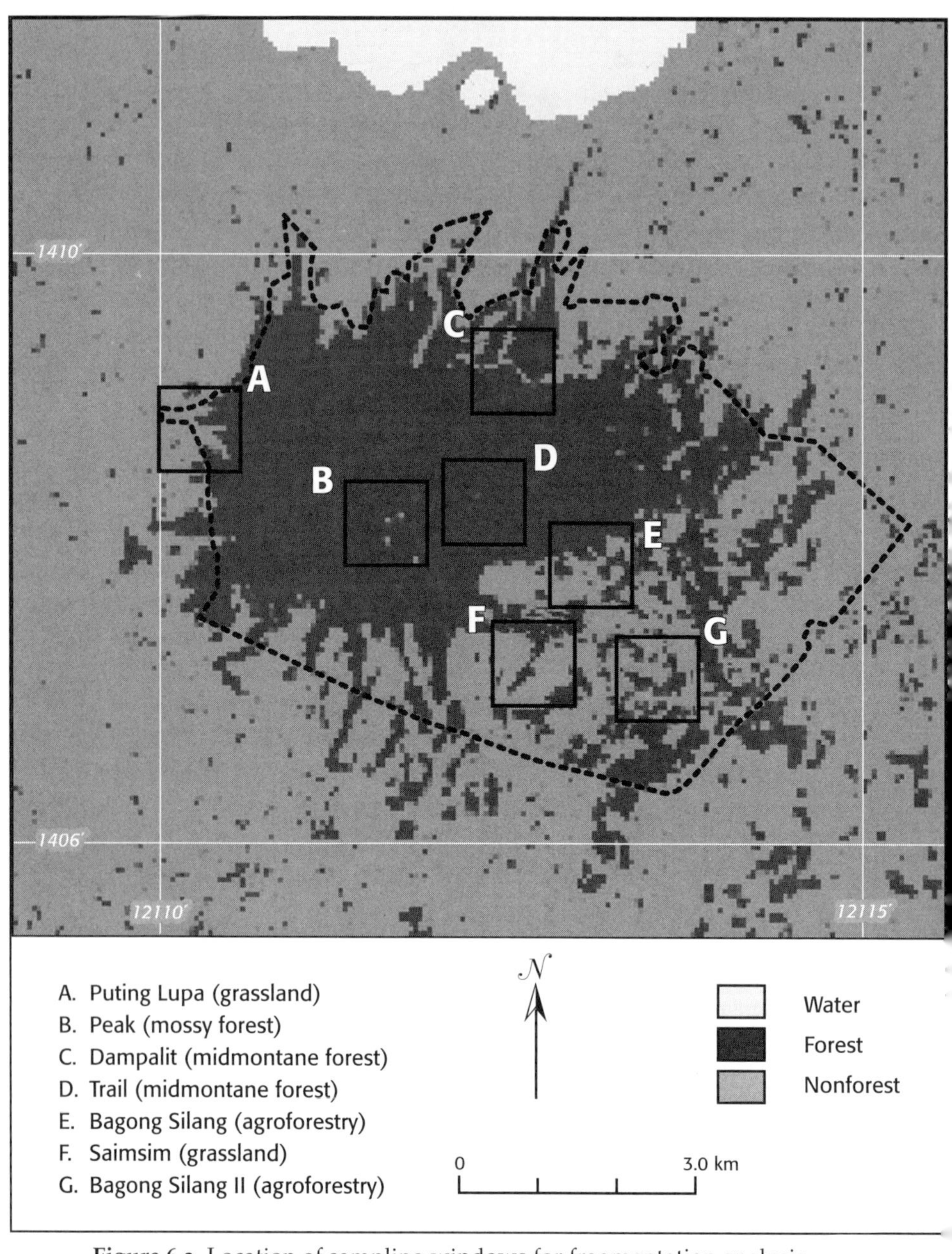

Figure 6.3 Location of sampling windows for fragmentation analysis

and Ticsay-Ruscoe 1997; Gruezo 1997; Gonzalez and Dans 1997; Gruezo and Gonzalez 1997; Francisco 1997) must be representative of the sampling windows. Figure 6.3 shows the location of each sampling window in the study site.

For each sampling window, the patch density and quadrat coefficients were calculated using the QUADRAT module of the IDRISI GIS package. To compute the fractal dimension, the nonforest polygons in the sampling windows were first determined and counted using the GROUP module of IDRISI. The corresponding areas and perimeters of the nonforest polygons were then calculated, using the AREA and PERIMETER modules. These values were exported to a spreadsheet, where the logarithms of the perimeters and the logarithms of the square roots of the areas were computed. The fractal dimensions were obtained by regressing the logarithms and calculating the regression coefficients for each of the seven sampling windows. These were used with the corresponding densities and quadrat coefficients to compute the fragmentation indices for each sampling window. The results are summarized in table 6.1.

Table 6.1 *Summary of fractal dimension (D), patch density (d), quadrat coefficient (Q), fragmentation index (f), biodiversity indices, and faunal endemicity*

	D	d	Q	f	Flora (H')	Fauna (H')	Endemism
Midmontane	1	0.00346	0.9974	0.00173	3.65083	4.244	0.67568
Mossy forest	1.51196	0.01384	0.98701	0.01053	3.56972	4.115	0.72973
Dampalit	1.25627	0.09537	0.90544	0.06287	3.73695	4.569	0.67213
Bagong Silang	1.38059	0.51515	0.48528	0.47884	3.22453	4.569	0.5625
Puting Lupa	1.29811	0.60963	0.39072	0.56903	1.35662	3.647	0.18182
Saimsim	1.21492	0.72371	0.27654	0.68877	1.34965		
Bagong Silang-Silang	1.43567	0.67474	0.32554	0.7308	3.17212		

Results and Discussion

The Nature of the Landscape Fragmentation and Its Proposed Index

Figure 6.4 shows the patchiness of the seven sample sites with their corresponding locations, ecotypes, fragmentation indices, and coefficients of the model components. The sample sites are arranged in order of increasing fragmentation from left to right and top to bottom. Because we are dealing with nonforest patches in a forested landscape, the forest appears black (class = 0) and the nonforest patches appear white (class = 1). The first three sampling windows, corresponding to the natural forest, show very few patches and correspondingly low fragmentation indices. The following four sampling windows, corresponding to grasslands and agroforestry areas, show greater fragmented densities, patch shape complexities, patch distribution regularities, and consequently higher fragmentation indices.

There were two instances in which the sampling windows yielded only two polygons each, the minimum number needed to derive the fractal dimension. These are for sites 5 and 7, with standard errors equal to zero in the regression analysis, indicating perfect fit. This is such since in each case there were only two points with which to estimate the regression line (and no other point not on the line). In both instances, one patch is a very large and complex polygon while the other is only one pixel in size. Such conditions are far from ideal for calculating the fractal dimension, which would be more accurate if there were more polygons of relatively equal sizes and complexities. The two instances mentioned will bias the fractal dimension to values closer to 1 and suggest that a larger sample window might have been more appropriate. However, we must also consider whether our biodiversity data would be representative of a larger sampling window. A compromise must therefore be reached in situations such as these between accurate estimations of patch complexity and true representativeness of biodiversity measures. In such cases, it seems wiser to bias the fractal dimension rather than compromise the level of significance of the biodiversity indices because the discussion to follow shows that the fractal dimension has

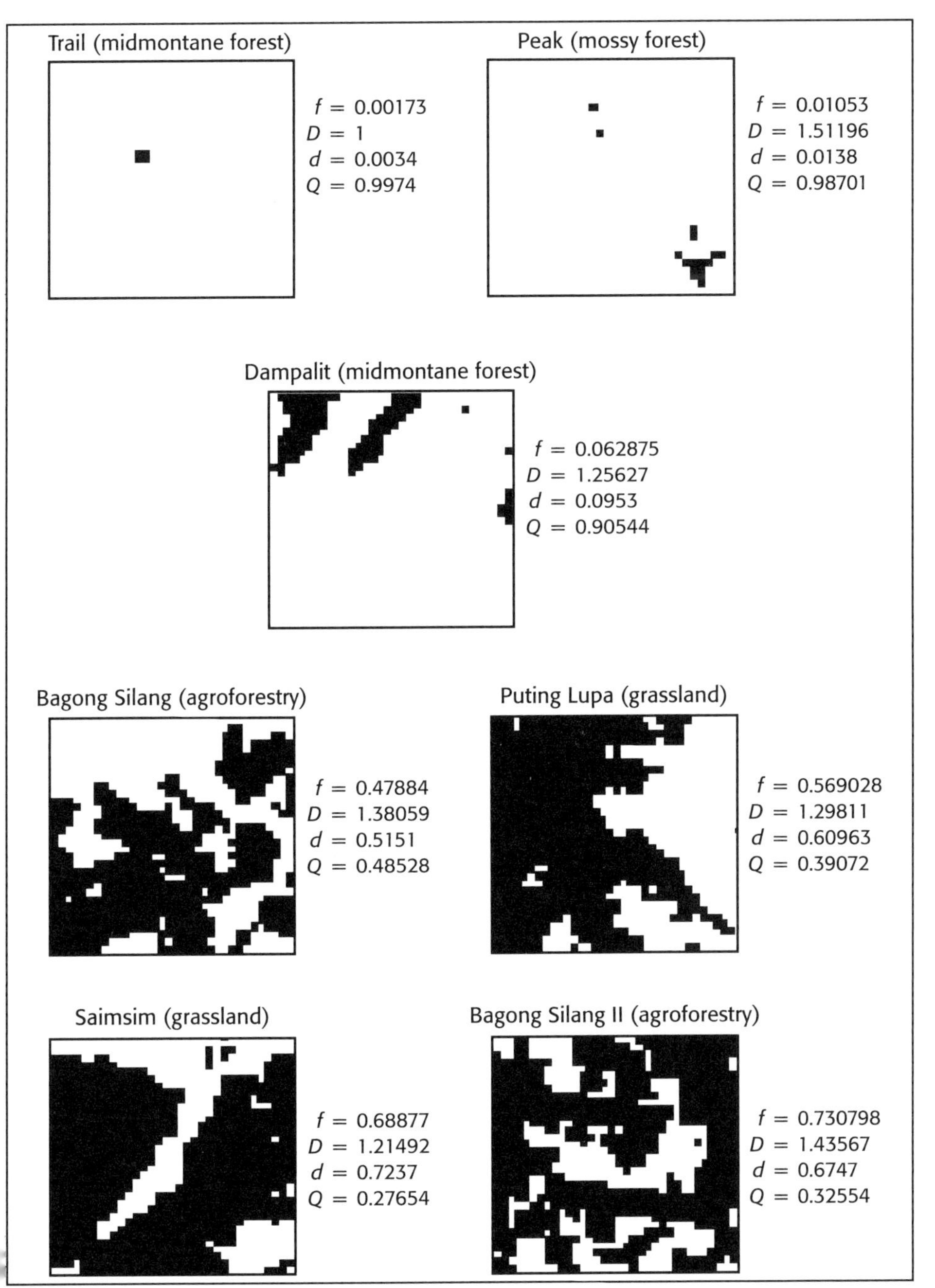

Figure 6.4 Sampling windows with locale, ecotype, fragmentation index (f), fractal dimension (D), patch density (d), and quadrat coefficient (Q)

less influence on the fragmentation index than do the other components of the model.

The sample site in one of the midmontane forest areas (site 1) had only one square patch in it, which presents computational problems. It is impossible to compute the fractal dimension of only one polygon, as we would need two or more polygons for the regression. This single polygon was assigned a fractal dimension of $D = 1$, such as we would have for squares and circles. This was an opportunity to test the use of the compactness ratio in lieu of the fractal dimension to estimate patch shape complexity, approximating $D = 1$ with $(2 - c)$ where $c = 0.886154$ calculated using the CRATIO module. But the corresponding absolute change in the fragmentation index was only 0.0001972, which at most only affected the fourth decimal place in the results of all the regression analyses involving the fragmentation index. Thus, the overall impact of the change in estimating the patch shape parameter in this one site was insignificant.

Regression Analyses

COMPONENTS OF THE MODEL. Because patch density and the quadrat coefficient are inversely related, they have almost equivalent coefficients and are strongly (see table 6.2, $r^2 = 0.9944$) and significantly ($p = 0.0000465$) related to the fragmentation index, although the relationship is direct in the case of patch density and inverse in the case of the quadrat coefficient. Thus, patch density and distribution have a high impact on the fragmentation index, as we would expect. In contrast and as mentioned earlier, the relationship between the fractal dimension and the fragmentation index is weak ($r^2 = 0.2365$) and insignificant ($p = 0.6095$). This justifies our earlier decision to compromise the estimate of patch shape complexity rather than compromise the biodiversity measures of the sample. It also reinforces our belief that as long as the density of the patches remains relatively unchanged over a landscape we can increase or decrease the size of the sample window to improve data representation within the area without too much regard for the complexity of the shapes of the patches. So the decision as to the size of the sample windows should be guided by how well the windows represent the populations within them, rather than the spatial characteristics of the

Table 6.2 *Summary regression coefficients and probabilities*

	r^2	P		
$f \times D$	0.055959	0.60955062	*	
$f \times d$	0.9888591	0.00000447	**	
$f \times Q$	0.988863	0.00000446	**	
$f \times (D, d, Q)$	0.9945032	0.00069072	**	
$d \times Q$	1	5.39E–24	**	m = -1.00086

*Significant at the 10 percent level.
**Significant at the 5 percent level.

patches themselves. Needless to say, the size of the windows must be more or less uniform within a single study to rule out any biases that varying sample window sizes may introduce

Overall, the fragmentation index is highly (r^2=0.9972) and significantly related (p = 0.00069) to its components if these are taken together. This speaks well for the robustness of the model for estimating patchiness in the landscape in that the degree of landscape patchiness can be well explained in terms of patch density, distribution, and shape complexity.

Patch density is very significantly (p = 5.39E-24) but inversely (m =-1.00086) related to the quadrat coefficient. The perfect relationship (r^2 = 1) is in fact expected, as the quadrat coefficient is the ratio of the patch density mean to its variance. This perfect relationship, however, raises the possibility of autocorrelation between the two components of the model. An alternative model was thus investigated, by omitting patch density, in the hope that the quadrat coefficient and the fractal dimension would account for the variation in patchiness. Table 6.3 shows the corresponding alternative fragmentation indices (f') and the results of the regression with the original fragmentation indices (f). Although the resulting association was fairly high (r^2=0.6318), it was not at all significant (p = 0.1277); thus, the alternative index does not estimate the degree of patchiness as well as the original model does. Also the alternative model is not as well behaved, as its limits are wider.

Table 6.3 *Fragmentation index (f) and alternative index (f′) of the seven sampling sites*

Locale	f	f'	
Midmontane	0.00173225	0.001929459	$f = (D^*d)/(1+Q)$
Mossy forest	0.01053193	0.760922688	
Dampalit	0.06287489	0.087271378	$f' = D/(1+Q)$
Bagong Silang	0.47884069	0.929513404	
Puting Lupa	0.56902784	0.933404803	$r^2 = 0.6318$
Saimsim	0.68877327	0.951728147	$p = 0.1277$ n.s.
Bagong Silang-Silang	0.73079827	1.083081284	

The fractal dimension is insignificantly (table 6.4, $p = 0.6497$) and poorly ($r^2 = 0.2109$) related to either the patch density or the quadrat coefficient. Again this is to be expected, as you can have infinitely many shapes for patches even if their density is held constant. Putting it another way, because patch density is the ratio of the number of patch pixels to the total number of pixels in the window, there are infinitely many ways to rearrange a given number of pixels in the window (shapewise and distributionwise) without changing the density. As for the quadrat coefficient (distribution) and the fractal dimension (shape complexity), given a fixed number of pixels in a fixed number of patches (constant density), you can move the patches around and change their distribution in space but still retain their shapes, or you can change their shapes without moving their relative locations.

Table 6.4 *Summary regression coefficients and probabilities for fractal dimension (D) with patch density (d) and quadrat coefficient (Q)*

	r^2	p	
$D \times d$	0.0445091	0.64976	n.s.
$D \times Q$	0.0445101	0.64976	n.s.
$D \times (d, Q)$	0.0514443	0.89976	n.s.

Note: n.s = not significant

This is also true when we take patch density and distribution together and relate it to shape complexity. The inverse relationships (m_1 = -1562.05 and m_2 = -1560.81) between the fractal dimension on the one hand and patch density and distribution on the other are poor ($r^2 = 0.2268$) and very insignificant ($p = 0.8997$). This means that, given a fixed shape for the patches, you can add to or deduct pixels from them and even move them around without changing their relative shapes. Or, given a fixed number of pixels and fixed positions of the patches in space, if you rearrange the contiguous pixels within each patch you can actually change the patch shapes without changing their densities or distributions.

BIODIVERSITY INDICES AND OTHER RATIOS. The Shannon-Weiner indices of flora and fauna diversity, as well as the ratio of endemic to total faunal species in the other studies done in the Mount Makiling Forest Reserve in the Philippines (see Sajise and Ticsay-Ruscoe 1997; Gruezo 1997; Gonzalez and Dans 1997; Gruezo and Gonzalez 1997; Francisco 1997) were regressed with their corresponding fragmentation indices. The results are presented in table 6.5. There were seven sites for the floral studies and five sites for the faunal studies. The relationships shown are quite loose, with 10 percent being the best level of significance attainable. The relationships with faunal endemism were the highest ($r^2 = 0.8547$) followed by floral diversity with five samples ($r^2 = 0.8121$), and then with seven samples ($r^2 = 0.6976$). The relationship with faunal diversity is very poor ($r^2 = 0.3508$) and insignificant ($p = 0.5626$). The poorness of fit means that the relationships between fragmentation, on the one hand, and biodiversity and endemism on the other are not linear, although an

Table 6.5 *Summary of regression coefficients and probabilities for fragmentation index and measures of biodiversity and faunal endemicity*

	N	r^2	p	
Flora × f (7 sites)	7	0.486769	0.081352729	*
Flora × f (5 sites)	5	0.65953	0.094959616	*
Fauna × f	5	0.123088	0.562638885	n.s.
Faunal endemism × f	5	0.730665	0.064958768	*

Note: *Significant at the 10 percent level, n.s. = not significant.

inverse trend is apparent. Investigating the behaviors of their graphs may reveal more insights into their relationships than do the regressions.

Figure 6.5 shows the relationship between floral and faunal biodiversity and fragmentation for the five sample plots. Here we see the familiar inverted U-shaped curve skewed to the left. As patchiness increases in the natural forest, biodiversity gradually increases as well. But further increases in fragmentation in the more human-influenced areas bring biodiversity down drastically. This relationship perhaps points to a threshold level for biodiversity with respect to patchiness. The graph for endemism and fragmentation (fig. 6.6) shows that as we move from the natural forest to the grasslands, endemism decreases geometrically. Within the midmontane forests, faunal endemism is relatively stable even when patchiness varies. But within the mossy forest endemism seems to be enhanced by

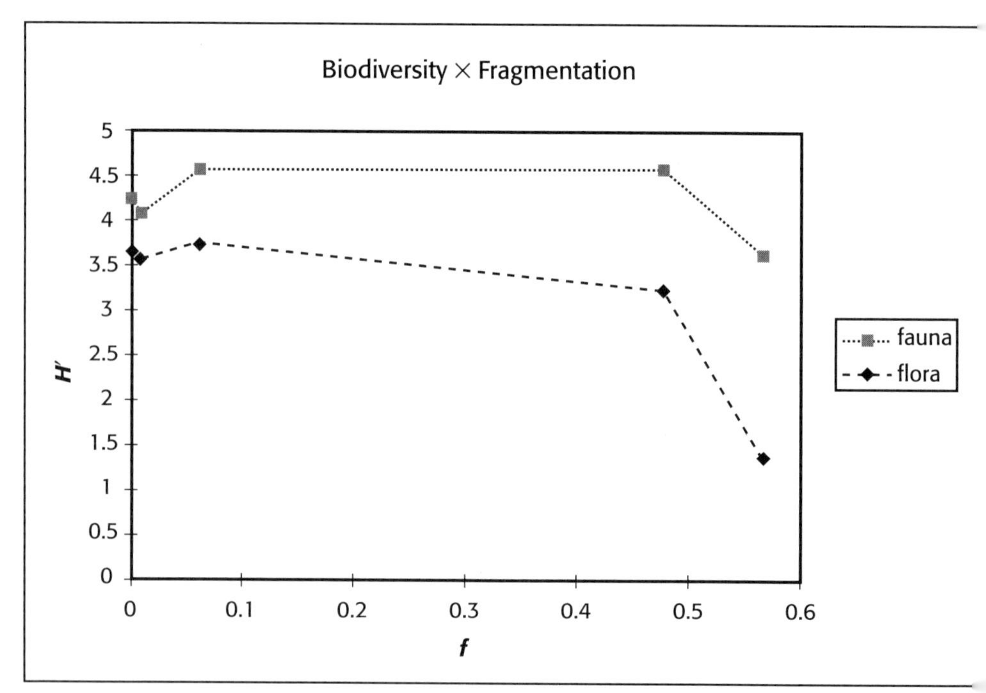

Figure 6.5 Biodiversity and fragmentation (five sites)

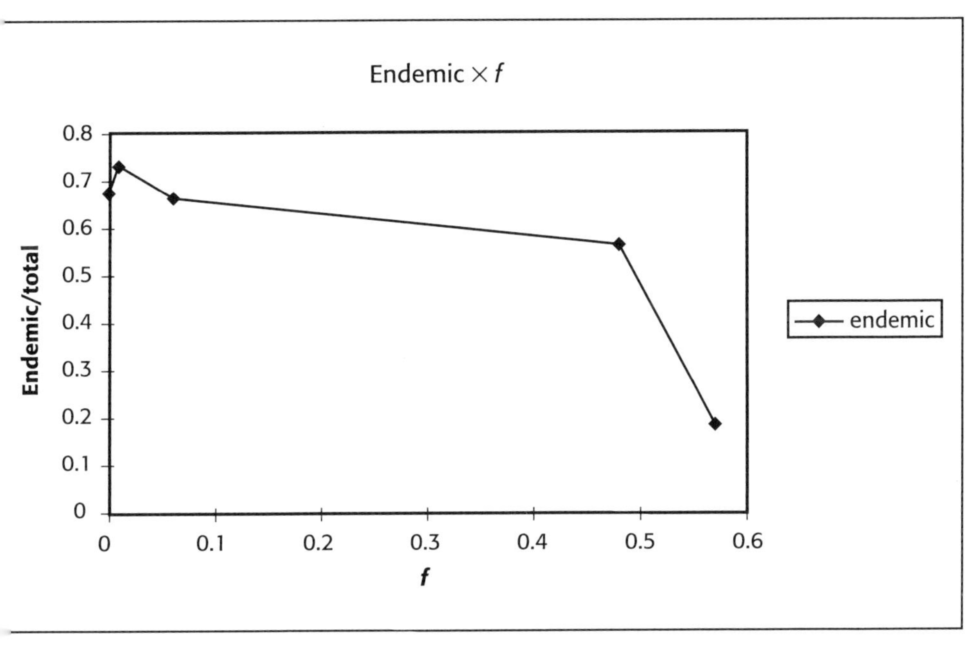

Figure 6.6 Endemicity and fragmentation

some factor other than fragmentation, possibly altitude or isolation, as endemism in the mossy forest is higher even though the patchiness index of the mossy forest lies between that of the midmontane forest sites. Generally, though, endemism dramatically drops as forest cover is reduced. Figure 6.7 shows the relationship between floral diversity and patchiness in the seven sites. The graph shows a wider range of patchiness in the agroforestry areas than in the grasslands, yet the biodiversity levels in the agroforestry areas are consistently higher than in the grasslands. So there must be other factors involved in maintaining and enhancing biodiversity levels than just forest versus grassland cover. However, although biodiversity levels may be maintained in the grasslands and agroforestry areas despite a wide variation in patchiness, endemicity is dramatically lower in these two areas. It is apparent that nonendemic species are replacing endemic ones even when overall measures of diversity are being maintained.

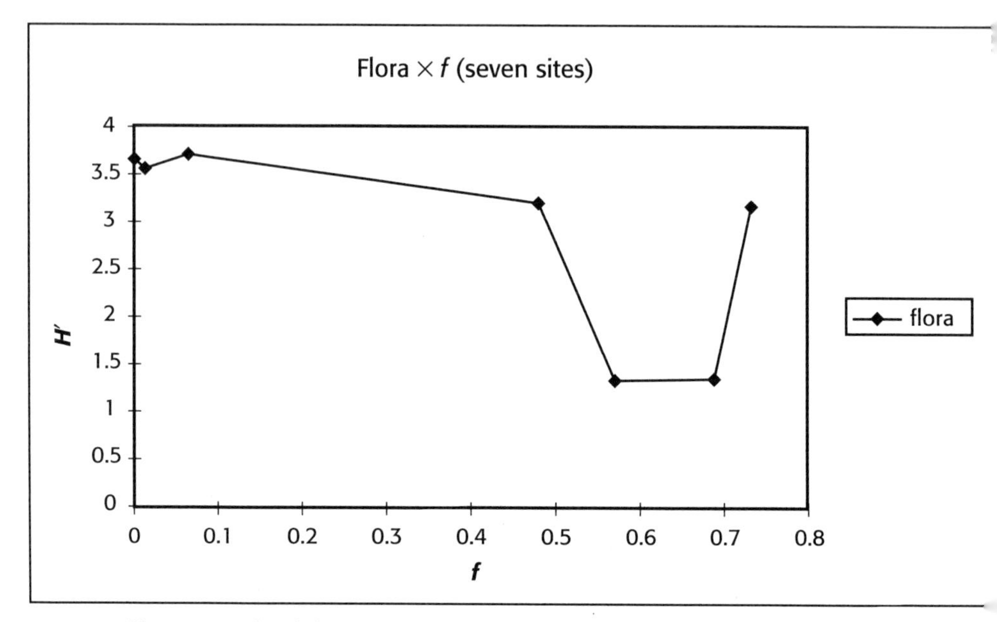

Figure 6.7 Floral diversity versus fragmentation index

Conclusion

Model Behavior

Our analysis showed that patchiness increases as patch density and shape complexity increase and patch clustering decreases. All values for the quadrat coefficient and the fragmentation index remained within the [0,1) range. It would have been interesting to see how the model behaved if the patches were clustered, but this opportunity was not afforded during this study. Due to the nature of the patches, such an investigation would entail a larger sampling window for this study site, and thus more sampling points, or a different study site with more clustered patches.

Another issue that would be worth looking into is whether the model is exhaustive and mutually exclusive. Putting it another way, since the fragmentation index for the nonforest patches is bounded within [0,1), if we subtract this from 1, do we get the corresponding fragmentation index for forest patches? If so, the implications would be immense. For example, if we suspected a bias in the estimate of any of the model parameters we could compute for the complementary index and subtract this from 1 to get the desired fragmentation index. It would be simple in the case of the present study to reverse the point of view and compute the fragmentation indices of forest patches to see if together with the complementary indices they added up to 1 for each sample site. Unfortunately, even if they did add up to 1 this still would not be sufficient to prove exhaustivity nor mutual exclusivity.

Policy Implications

The results from the regression analysis and the graphs concerning patchiness, biodiversity, and endemicity point out two things. First, there is some positive association between patchiness and biodiversity. Increasing patchiness in the natural forest landscape increases biodiversity slightly, but beyond a certain point increasing patchiness in the landscape results in dramatic decreases in biodiversity. It would be interesting to investigate whether or not a threshold value really does exist with the tools we now have in hand. Such a study should also consider the size and proximity of the natural reservoir (in this case, the forest) from which resources can be drawn to enhance or maintain diversity. The maintenance of relatively high levels of flora in the agroforestry areas, in the face of high levels of habitat fragmentation, shows that human perturbation may contribute to the maintenance of high levels of diversity (though at the expense of faunal endemism). Relatively high levels of plant diversity are also maintained in the grasslands even with high levels of fragmentation, possibly underscoring an anthropogenic factor for plant diversity maintenance in the grasslands, though probably not as direct as in the agroforestry areas.

Second, the data on endemism show that when faunal biodiversity increases as patchiness is increased endemism is actually reduced. This agrees with other findings on species loss and fragmentation (Kruess and Tscharntke 1994). However, we should also keep in mind that some context-specific factors may have been at work here, as no large grazing mammals that would normally be at an advantage in patchy landscapes were observed during the study periods.

For the planning and management of forest reserves with resident settlers in a manner designed to maximize biodiversity conservation, the results presented here reinforce the concept of leaving the major portion of the forest reserve intact and regularly shaped to act as a biological resource pool. Clearings should be distributed in the buffer zone in as sparse, discontiguous, and clustered a manner as possible. Clustering the clearings will leave more space untouched for wildlife habitat. Clearing large, contiguous tracts of forest should be avoided. Preferably, the clearings should be regular in shape, square or circular, as opposed to long and thin or other complex shapes. However, our results suggest that the density and distribution of the patches or clearings are more important than their shape.

Notes

1 One may argue about the merits of using the mean of the compactness ratios of the polygons to deal with a group of them. Unfortunately, this is beyond the scope of this chapter, but such an investigation might be pursued if the proposed index stirs enough interest.

2 Initially the IDRISI version 4.1 that ran under DOS was used; later the IDRISI for Windows version 1.0 was used.

Part IV: Regional and Global Perspectives

7

Biodiversity Conservation in the Mount Makiling Forest Reserve, Laguna, Luzon

PERCY E. SAJISE, MARILIZA V. TICSAY, WILLIAM SM. GRUEZO, JUAN CARLOS T. GONZALEZ, ANDRES TOMAS DANS, HERMINIA A. FRANCISCO, CLEOFE S. TORRES, DANTE K. VERGARA, AND VERNON VELASCO

THE PHILIPPINES IS CONSIDERED to have one of the highest concentrations of biodiversity in the world. It has a high endemicity of flora and fauna, many of which are endemic subspecies unique to each of the over seven thousand islands. As a result of rapid rates of deforestation and rural transformation, however, this biodiversity has declined precipitously, so that today the Philippines is considered to be one of the "hot spots" for biodiversity degradation and conservation.

This chapter describes the results of a study conducted by a multidisciplinary team of experts in the Mount Makiling Forest Reserve (MFR) from 1994 to 1996. The study espoused a new approach to conservation, in which the focus was not directly on studying the preservation of biodiversity but on the preservation of the bio-social conditions that promote the conservation of biodiversity. The research was less an attempt to gather new data than to reinterpret existing data, to make new linkages among existing data, and to analyze the relationships among several hierarchical levels, namely, farm, ecosystem, and landscape. The study also aimed to generate new methodologies for the rapid assessment of biodiversity.

Methodologies for linking the results of the study to conservation policy and management were also an important concern.

The discussion begins with a floral diversity profile of the different ecosystems in the MFR. The utility of floral species as indicators is also discussed. The next section, on faunal diversity, presents an analysis of the field data on the diversity of amphibians, reptiles, birds, and mammals on the mountain and discusses the impact of habitat disturbance on this diversity. It also discusses the implications of community structure for diversity, the status of threatened species on Mount Makiling, and the sociocultural importance of fauna. The final section, on socioeconomic determinants of crop diversity, discusses the relationship between biodiversity levels at the farm or agroecosystem level and the socioeconomic attributes of the farmers using a rapid assessment methodology developed especially for the study.

The MFR was selected for this study because it is one of the few remaining areas in the Philippines with a relatively large portion of intact natural forest containing many diverse species of fauna and flora. At the same time, it has a long history of human activity and encroachment resulting in conversion of the forest to other land use systems, which presents us with varied levels of patchiness, a wide spectrum of biodiversity levels, and a continuum of nature and society interactions.

Methodology

The methods used to measure biodiversity at the farm level included standard indices of ecological biodiversity, namely, the Shannon-Weiner index of biodiversity (H′), the Menhinick index of species richness, and Pielou's evenness index (computed based on the number of species and individuals per species derived from plot sampling for floral species). The research team also employed rapid biodiversity appraisal (RBA) methodologies to assess species distribution and abundance, including a participatory transect walk (which uses indicator species to assess biodiversity and consists of species identification and a record of the total number of species found in a particular sampling area as opposed to the number of individuals per species in a given sampling space), and the crop diversity index of qualitative variation (CD-IQV) which integrates bio-social information and the total number of species found.

The Study Area

Mount Makiling is an isolated, extinct volcanic cone located 65 km southeast of Manila in the south-central part of Luzon Island (lat. 14°8′ N and long. 121°12′ E; see fig. 7.1). The crater is not readily visible, but its rim may be traced from individual peaks along worn, deep channels and steep slopes. The topography of the lower slopes consists of broad, radiating, well-drained ridges separated by narrow valleys. The highest peak reaches only 1,130 m asl. Its volcanic origins are evident in the many hot springs and fumaroles. The reserve has an area of 4,244 ha, including portions of two provinces, Laguna and Batangas (Miranda 1987). It is situated on the southwestern Luzon volcanic plate, considered to be the "meeting point" between the two distinct faunal regions of northern and southern Luzon, running from the tail end of the Sierra Madre of northeastern Luzon to the upper limits of the Bicol Peninsula of southern Luzon.

Mount Makiling was one of the first national parks established in the Philippines. The Bureau of Forestry first classified it as a forest reserve and a field laboratory of the University of the Philippines, Los Baños (UPLB), Forestry School in 1910 under Presidential Proclamation 106. Proclamation 552 of 1933 proclaimed it a national park. In 1993, it was reclassified as a forest reserve through Republic Act 6,967. The Makiling Forest Reserve is public land placed in trust under management of UPLB, which houses the country's foremost forestry and agricultural school. The reserve is utilized by the university as a natural laboratory for research, extension, and instruction in forestry and agroforestry. The reserve lands also are used for the Makiling Botanical Garden, the UPLB Museum of Natural History, the National Arts Center, and the national campsite of the Boy Scouts and Girl Scouts of the Philippines. The reserve also functions as a vital watershed for the surrounding municipalities in the provinces of Laguna and Batangas (WBL 1994).

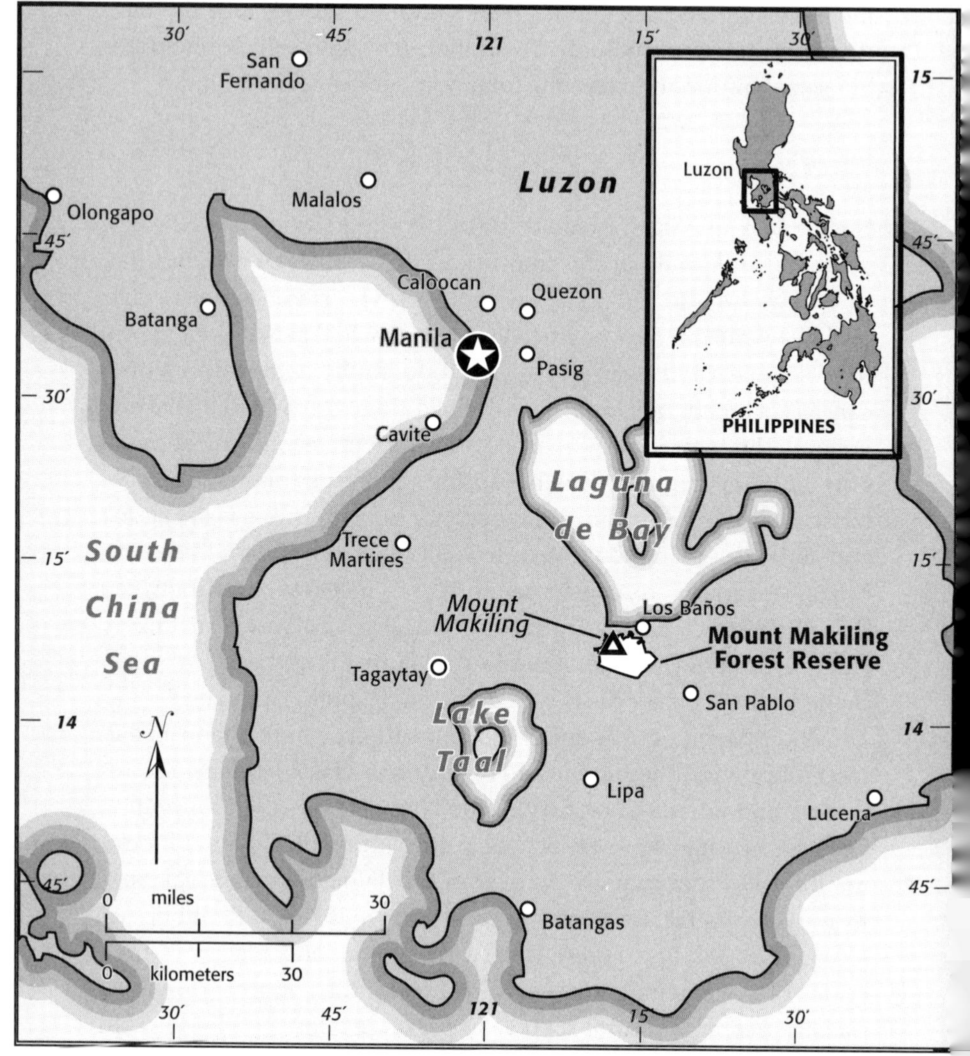

Figure 7.1 The Mount Makiling Forest Reserve, Los Baños, Philippines

Historical records of human occupancy on Mount Makiling date as far back as 1898, more than ten years before the UPLB School of Agriculture was established. Settlement increased during World War II, when lowland families took refuge on the mountain, opening large tracts for *kaingin* (shifting cultivation) in order to survive. What began with a few families has since grown. Today six distinct clans dominate the various settled areas of the reserve. A recent census (Dizon 1992) showed that there are 1,100 land claimants in the reserve, 25 percent of whom actually reside within it. Recent increases in agricultural lands within the reserve from 528 ha to around 1,900 ha, at the time of the 1992 farm inventory conducted by the Institute of Forest Conservation (IFC), attests to the potential role played by the settlers in the conversion of the reserve's forests (Cruz, Francisco, and Torres 1991).

Over the years UPLB has made a number of efforts to address the problem of human occupation of the reserve, ranging from expulsion to cooperation. In the 1990s, the university entered into a partnership with the local communities to jointly protect the remaining forest cover. Although this partnership has not completely halted encroachment into the reserve, it seems to have slowed it. A threat still remains, however, from the population growth of the current forest occupants. The population growth rate is quite high and is expected to remain so given the large percentage of young people living in the reserve (Francisco, Mallion, and Sumalde 1992). Furthermore, the only remaining area for further agricultural expansion is brushland and the forest zone.

Floral Diversity

The vegetation on Mount Makiling was formerly dipterocarp forest from 100 to 600 m asl. However, due to historic anthropogenic disturbances (notably, shifting cultivation and timber poaching), much of the lowland forest cover has been lost. The introduction of exotic trees for reforestation and orchards on the lower slopes has further transformed the lowland forest below 300 m to a mosaic of secondary growth, orchards, and grasslands.

In order to study all of the vegetation types present, the project established research sites within each of Mount Makiling's four different types of habitats, ranging along disturbance and elevational gradients (from 350 m up to the second peak at 1,110 m asl) within the six hydrological zones of the Makiling watershed. At each study site, extensive vegetation sampling was undertaken, consisting of identification and description of major plant associations, a detailed listing of floristic composition, a study of their relationships, and distribution mapping. In addition, an ethnobotanical survey was undertaken with the assistance of key informants, field guides, and community members, to identify the economic, social, and other functions of floral resources identified during the study.

The Mossy Forest

The mossy or cloud forest zone begins to manifest itself at an altitude of 900 m and has its peak of development in the vicinity of Makiling's three peaks of slightly varying heights (1,085, 1,090, and 1,109 m). With the exception of a well-trodden trail from the Mudsprings area to peak 2, the mossy forest zone is the least disturbed portion of Mount Makiling. The mossy forest zone is characterized by the presence of a dense growth of epiphytic bryophytes (liverworts and mosses) on the crooked, moist trunks of dwarfed trees. Overall the floral diversity in this zone ranked third among Makiling's four ecosystems, with a total of ninety-five species distributed among seventy-three genera and forty-eight families. Of these, twenty-three species are considered unique to the summit area of Mount Makiling.

The Dipterocarp Midmontane Forest

The dipterocarp midmontane forest zone of Makiling starts at an elevation of about 100 m and has its upper boundary limit at 900 m. At the latter elevation, this zone merges with the mossy forest. Also known as the "high forest" due to the historic dominance of very tall, timber- producing species belonging to the family Dipterocarpaceae, this forest type exhibits some state of deciduousness, particularly during the dry season. The dipterocarp forest persists today outside two distinct enclaves of grasslands, several large abandoned agricultural clearings, and scattered, more or less permanent farms.

Above 500 m asl, another subzone begins, marked by the presence of tree species belonging to families other than Dipterocarpaceae such as Myrtaceae, Sapotaceae, Moraceae, Saurariaceae, Annonaceae, Sterculiaceae, and Tiliaceae. Two distinct layers are formed by these trees: the upper (A) story is composed of large trees reaching a height of 25 to 35 m, while the lower (B) story has shrubs and trees that grow no higher than 10 m. This forest zone extends up to 900 m and has been classified as midmontane forest or simply rain forest. Here we do not distinguish between the dipterocarp forest and the midmontane forest due to the rarity of the dipterocarps. Instead, the two vegetation types have been combined to form what is referred to here as the "dipterocarp midmontane" forest zone. At present, the most dominant tree species on these sites is *Diplodiscus paniculatus* of the family Tiliaceae, locally known as *balobo*. Its abundance in this zone is attributed to the fact that birds and mammals and even humans eat its fruits. These dispersal agents widely and effectively distribute its seeds.

Project researchers identified in this zone 373 species distributed among 263 genera and 95 families. Although this zone has the highest number of species of any zone studied on the mountain, its diversity index ranks second to that of the agroforestry zone. Of the 373 species recorded here, 64 are endemics and 309 are native or indigenous. One introduced timber species, large-leaf mahogany *(Sweitenia macrophylla)*, is now successfully growing wild in nearly pure stands in isolated spots within this zone. The relative plant density in this zone ranges from 142 to 2,824 individuals per study plot (area = 0.8 ha), with species richness ranging from 38 to 102.

The Grassland Ecozone

There are only two localities where grasslands are found in the Mount Makiling Forest Reserve, namely, Puting Lupa (Calamba) and Sitio Nagtalaok-Saimsim, Bagong Silang (Los Baños). In both localities, it is apparent that maintenance of the grassland ecozone is primarily a function of human activity, especially wildfire. These sites are thus anthropogenic in origin and nature.

In Puting Lupa, *Saccharum spontaneum (talahib)* dominates on ridge slopes and in shallow depressions. Small patches of *Imperata cylindrica (cogon)* are found on ridge tops and along trails leading to the dipterocarp midmontane forest zone. In addition to these two common grass species, two other grasses were frequently noted, namely, *Apluda mutica* and *Pennisetum purpureum*. The latter was purposely introduced into this site for fodder production and has become naturalized. Several species of ground orchids *(Geodorum nutans, Habenaria malintana, and Nervilia discolor)* were also found growing between culms of grasses.

The grassland zone in Nagtalaok-Saimsim yielded fifteen species (S) of plants. The dominant grass species in this grassland is *S. spontaneum*, while *I. cylindrica* is apparently absent. However, four other grass species *(Pennisetum purpureum, Rottboellia exaltata, Setaria palmifolia*, and *Paspalum conjugatum)* were found in association with *S. spontaneum*. The grassland ecozone had the lowest aggregate diversity index and total number of species ($S = 44$) among the mountain's four ecozones.

The Agroecosystem

The agroforestry zone on Mount Makiling is largely a mosaic of land uses: (1) coconut plantations intercropped with coffee, bananas, and other commercial fruit trees; (2) citrus orchards bounded by second-growth forests; (3) citrus orchards enclosed by large tracts of coconut plantations; (4) old coffee plantations (partly abandoned) adjacent to or swamped by mature secondary forests; (5) mixed fruit plantations with aroids, pineapple, and vegetables; and (6) newly opened swidden fields abutting mature secondary forests. Because of the diversity inherent in this category, different sites or localities classified as agroforest exhibited very different levels of floral diversity.

A total of 367 species in 290 genera of plants were recorded from 15 study plots established within the agroforestry zone. This total can be broken down into 45 crop species (18 annuals and 27 perennials), 79 weedy species, and 243 forest species. The crop species component of the agroforestry zone can be further classified into 24 fruit tree species, 15 vegetable species, 7 root crop species, 4 ornamentals, 3 beverage species, 3 condiments/spice species, 3 species for soil erosion control, 2 species each for fuelwood and construction materials, and one species each for masticatory use, cover crops, dyes, and medicinal use.

Analysis of the weedy component of the agroforestry zone reveals that many species have some productive alternative function. Examples of these alternative functions include (1) fodder, (2) vegetables, (3) cover crop or green fertilizer, (4) edible fruit, and (5) herbal medicine. Undoubtedly, other species also perform ecological functions such as sources of nectar for insects, birds, and mammals; fruits and seeds for frugivores; habitats and sources of food for the insects that are common in swidden fields and nearby secondary forests; and soil erosion control. A number of forest species present in the agroforestry zone also perform useful functions as sources of (1) edible fruits, (2) vegetables, (3) pulp and fibers, and (4) ornamentals. Many forest species present in the agroforestry zone also function as pioneer species in ecological succession.

In terms of total number of species identified (S = 367), the agroforestry zone ranked second behind the dipterocarp midmontane forest zone. Species abundance was also comparatively high ranging from 405 to 2,705 individuals per study plot.

A comparison of aggregate zonal diversity indices of the four ecosystem zones in the MFR is given in table 7.1. The agroforestry zone has the highest aggregate index (H′), followed by the dipterocarp midmontane forest zone, then the mossy forest zone, and finally the grassland zone. In terms of species richness (R), however, the dipterocarp midmontane forest zone is slightly higher than the agroforestry zone. The mossy forest and grassland zones have relatively lower species richness. The slight difference in diversity indices between the dipterocarp midmontane forest and the agroforestry zone is due to the presence of a large number of introduced plant

Table 7.1 *Interecosystem zone comparison of species diversity in the Mount Makiling Forest Reserve*

Ecosystem Zone	N	S	Rank	H' Value	Rank
Mossy zone	731	95	3	3.56972	3
Grassland zone	5,242	44	4	1.66438	4
Dipterocarp midmontane zone	17,965	374	1	3.89713	2
Agroforestry zone	17,574	368	2	4.2869	1
Overall	41,512	682	—	—	—

Note: N = the total number of individuals for all species, S = the total number of species, and H' = biodiversity index. In terms of the total number of species, the dipterocarp midmontane zone ranks first (has the highest value), but in terms of biodiversity level, the agroforestry zone ranks first.

species in the agroforestry zone—primarily cultivated crops and their concomitant battery of weed species.

Faunal Diversity on Mount Makiling

Faunal data collection and cataloging went hand in hand with the floral analysis. A combination of recording methods was used, including (1) ocular and audio methods, (2) trapping/netting, (3) analysis of physical disturbance, and (4) ethnobiology. Data were gathered on the four higher vertebrate groups at ten study sites on Mount Makiling (in coordination with the floral component of the study). Faunal diversity at the ten sites was compared to assess differences and similarities between habitats and zones, particularly between forests and anthropogenic ecosystems.

Diversity of Amphibians

A total of eleven species of amphibians in three families were recorded (see appendix, table 7.1A). Of the eleven, nine are considered endemic to the Philippines. The other two are considered commensal species, which favor areas with human habitation. Four of the endemic species are found only in the Greater Luzon faunal region. These include Woodworth's frog *(Rana woodworthi)*, the Luzon narrow-mouthed frog *(Kaloula kalingensis)*, and two undescribed species. One of these two new species, the Luzon forest frog *(Platymantis* sp. nov A.) is is found only on Mount Makiling and Mount Banahaw, and the other, the Mount Makiling forest frog *(Platymantis* sp. nov B), is endemic only to the mountain (A.C. Alcala, personal communication 1996).

Ranid frogs (family Ranidae) predominate in the study area. The highest number of species belongs to the genus *(Rana)*, which is explained by the closeness of the study sites to water sources. A significant number of species also belong to the genus *Platymantis*. They are strictly forest species, do not require a nearby water source for survival, and are an important indicator of the humid microclimate of the forest. The dominant amphibian species recorded on Mount Makiling is the common forest frog *(Platymantis dorsalis)*, which was found at eight out of ten sites. The corrugated forest frog

(P. corrugatus) and the Luzon forest frog follow, with six and five sites, respectively. Endemicity was high at each site, as most of the frog species present were endemic.

The highest diversity of amphibians was found in the midmontane dipterocarp forest at 500 to 600 m asl, followed by the lowland dipterocarp forest, the mossy forest, and finally the midmontane dipterocarp forest at 400 to 500 m asl. All four of these sites have more or less undisturbed aquatic habitats and forest vegetation. The correlation of endemicity and diversity at these sites reflects the quality or health of the environment compared to the other study sites, which were more disturbed by human activity. Amphibians are important indicators of the condition of vegetation and the quality of water and can indeed be used to determine the extent of human disturbance of the landscape (Alcala et al. 1995).

Diversity of Reptiles

A total of nineteen reptiles (eleven lizards and eight snakes) in seven families were recorded at the ten study sites on Mount Makiling (see appendix, table 7.2A). There were twelve endemic Philippine species, two of which are endemic only to Luzon Island. The reptilian fauna was dominated by skinks (family Scincidae), represented by eight species, and Colubrid snakes (family Colubridae), represented by five species. The dominant species in terms of abundance were Cox's sphenomorphus *(Sphenomorphus coxi)* and Steere's sphenomorphus *(S. steerei)*, both observed at six sites, followed by Jagor's sphenomorphus *(S. jagori)*, observed at five sites. Although the current list of species for this study is comparatively high for any study area in the Philippines, an improvement in the surveying techniques for herpetofauna might yet add to it, as many reptiles are cryptic by nature and naturally occur at low population levels.

For the reptiles, the highest diversity indexes were found in the midmontane dipterocarp forest at 500 to 600 m asl and at 600 to 750 m asl, both sites lying between the mossy forest zone and the agroforestry zones at Barangay Bagong Silang. The data show that reptile diversity in the forest ecosystem was significantly higher than in the anthropogenic ecosystem, which harbored more commensal rather than indigenous species.

Diversity of Birds

A total of 109 bird species in 42 families were recorded at the ten study sites on Mount Makiling (see appendix, table 7.3A). Of this total, 49 species are endemic to the Philippines and 7 of these are endemic only to the Greater Luzon faunal region. Although endemicity at the species level was high, the number of endemic subspecies may prove to be even higher.

Transect counts for birds yielded high diversity indices for both the monoculture plantations and the agroforestry areas, although both represent anthropogenic sites situated near highly disturbed forests. Both of these sites had higher diversity indices than the forest sites (either mossy or midmontane dipterocarp forests). However, although the two anthropogenic sites had higher H′ values for avian diversity, it was the midmontane dipterocarp forest site at 500 to 600 m asl on the northeast side of Mount Makiling that yielded the most species (56). Further, although species diversity was found to be highest at these three sites (monocultural plantation, agroforestry zone and midmontane forest), endemicity was not. The mossy forest site had the highest endemicity at 78.3 percent. The midmontane dipterocarp forest at 450 to 500 m asl had 72.7 percent. This suggests that these forest sites, despite their low diversity, are vitally important as habitats for less tolerant endemic species. Diversity should not be measured only by total numbers and values but also by the composition of the species and their status in a particular ecosystem, including endemicity.

Diversity of Mammals

Project researchers recorded a total of thirty species of mammals on Mount Makiling, twelve of which are considered endemic to the Philippines (see appendix, table 7.4A). The majority of the mammalian fauna were bats and small mammals (such as rodents and shrews), seventeen of which were volant mammals. The bats included six frugivorous species (suborder Megachiroptera), which play an important ecological role in forests and agroecosystems as dispersal agents of tropical forest fruits and seeds, as pollinators, and as biological controls for many airborne pests and harmful insects.

Comparison Across Zones for All Faunal Groups

Analyses of biodiversity typically break down diversity indices by faunal groups to compare different sites, treating birds, mammals, and herpetofauna separately. In this study, however, an aggregated value for diversity was determined for all four vertebrate groups to assess faunal diversity across the different study sites.

The aggregate values show that diversity increases from grasslands (lowest value) to mossy forests, then the agroforestry zone, and finally the midmontane dipterocarp forests (highest value). The high diversity values in the dipterocarp forests are attributed to the presence of multiple niches created by the vertical layering of the forest (Gonzalez 1992). Tropical forests tend to harbor larger numbers of species than anthropogenic areas due to this greater vertical stratification. The comparatively high value found in the agroforestry areas, however, can be attributed to the fact that forest species that are intolerant of disturbance are replaced with colonizing species that favor it, and thus the overall number of species remains high. The relatively lesser diversity of species in the mossy forests can be attributed to elevation, as an increase in elevation generally means a decrease in diversity for most fauna and to a decrease in the stratification of the forest that reduces habitat variety.[1]

Analysis of faunal species richness (R) shows a general increase in values from the most anthropogenic areas to those with the least human disturbance. Thus, the values are the lowest in the agroforestry areas, higher in the grasslands, higher still in the midmontane dipterocarp forests, and highest in the mossy forests. Thus, diversity increases along a human disturbance gradient, with the least diversity found on sites with the most human disturbance and the highest diversity found on sites with the least human disturbance.

The resulting diversity indices aggregated from all faunal species recorded on Mount Makiling follows a pattern similar to that of the indices for avian diversity: the highest value (H′) was observed in the agroforestry zone, followed by the midmontane dipterocarp forests and the monocultural plantation. Thus, the anthropogenic ecosystems can harbor a diversity of species on a par with the forest ecosystem. Species diversity in the disturbed zones is heavily influenced by proximity to undisturbed zones; however this influence stems from the edge effect created by two converging areas such as forests and agroecosystems. The merger of two ecosystem types results in the mixing of two sets of fauna: tolerant forest species and colonizing nonforest species. In any case, quantitative measures of diversity cannot stand alone; they must also be analyzed in the light of the overall functioning of the ecosystem.

Community Structure and Faunal Diversity

In this study, greater diversity was found to be associated with maturity of vegetative succession (and height of the vegetative cover), which is attended by more distinct divisions of habitats and increasing niche diversity (Miranda 1987). The most significant determinant of the diversification of the forest habitats on Mount Makiling is vertical stratification. The vertical structure of the mature forest is divided into several definite layers (e.g., the emergent layer, main canopy, subcanopy or middle layer, understory, undergrowth or ground cover, forest floor, and subterranean soil), and different species occupy different layers. The midmontane dipterocarp forest is well stratified, and the birds that inhabit it show a definite vertical distribution from the top layer to the forest floor.

Layering of the vegetative cover is greatly influenced by elevation, which affects ambient temperature and humidity, which in turn affects the composition and growth of vegetation. The lowland humid forest contains the most layers and offers the most niches to fauna. As elevation increases and temperature decreases, vertical stratification is reduced and there is a reduction of diversity for some faunal groups (MacArthur and MacArthur 1961). This can be observed by comparing figures from the mossy forest site at the peak of Mount Makiling and the midmontane dipterocarp forest site nearest its base: the figures from the mossy forest show less diversity than those from the dipterocarp forests.

The multiplicities of crops in the mountain's agroecosystems produce stratification in vegetation that approximates the vertical layering found in the forest. For example, orchards contain a main canopy, coconuts provide an emergent layer, and cultivated crops along with weeds provide the understory.

The Sociocultural Importance of Fauna

Many of the settlers on Mount Makiling are familiar with large edible animals and birds such as deer, wild pig, hornbills, pigeons, doves, junglefowl, quail, civets, frogs, monitor lizards, and pythons. Other animals are recognized for particular uses, including the red junglefowl *(Gallus gallus)*, locally called *labuyo*, which is kept for use in cockfighting, and parrots such as the *colasisi (Loriculus philippensis)* and the *guaiabero (Bolbopsittacus lunulatus)*, which are caught alive in glue traps and sold as pets. Some commensal species are recognized because of the damage they cause to crops or human habitations. Examples are the oriental house rat *(Rattus tanezumi)* and the Asian house mouse *(Mus musculus)* and, because they favor human habitations, the colonizing commensal house gecko and the Eurasian tree sparrow *(Passer montanus)*. Others, such as snakes, are recognized because of the harm they can inflict (although only a small fraction of the total snake fauna in the Philippines is lethal to man). Rarely are fauna recognized as ecological indicators, although some residents do regard certain fauna as important vectors for pollination and dispersal based on their personal experience. For example, one informant reported that he had observed sunbirds pollinating the flowers of coconuts and garden plants around his house; other respondents claimed that an abundance of bats can be interpreted as a sign of an abundance of fruit-bearing trees nearby.

Key informants were also asked about the significance of particular wildlife species: deer *(Cervus mariannus)*, wild boar *(Sus* spp.), monkey *(Maraca fascicularis)*, bats *(Pteropus vampyrus* and *Ptenochyrus jagorii)*, and civet cats *(Paradoxurus philippinensis* and *Viverra tangalunga)*. The informants replied that these species are rarely observed in the more accessible forest and when sighted they are hunted for food. In spite of a number of laws and administrative orders prohibiting the hunting, trapping, catching, or killing any kind of game or nongame bird or taking the nests or eggs of any such bird in the forest reserve, disturbance of wildlife is still very evident. Commercial as well as sport hunters enter the reserve illegally to kill for food, trade, and trophies a number of species, most of which are vulnerable to extinction. And, although the local residents in Barangay Bagong Silang have begun to regulate the entry of outside poachers and wood collectors into the reserve, they themselves continue to harvest wildlife from the adjacent forest for food, additional income, and various sociocultural uses

Summary of Findings on Faunal Diversity

A total of 170 species of fauna (109 birds, 30 mammals, 19 reptiles, and 12 amphibians) were recorded at the ten study sites on Mount Makiling. This brings to a total of 311 the number of species of vertebrates known to occur on the mountain, including those previously recorded by other researchers. This total consists of 22 species of amphibians, 65 reptiles, 181 birds, and 45 mammals. About 135 species of the total are endemic. These are relatively high measures of diversity, almost rivaling the number of wildlife species found in the Sierra Madre in Cagayan and Isabela, in the northern Philippines (Danielsen et al. 1991). This is comparable to the diversity of species found in other national parks in southern Luzon such as Mount Isarog or Mount Kitanglad in Mindanao in the southern Philippines.

Project researchers observed more than half of the species previously recorded on Mount Makiling in addition to recording a number of new species. The project recorded almost half of the known amphibians occurring on Luzon Island, about 20 percent of the known reptiles, 45 percent of the known birds, and 30 percent of the known mammals. Nine threatened birds, four threatened and rare amphibians, three threatened and rare reptiles, and four threatened mammals were observed on Mount Makiling during the study. Such species are valuable indicators of habitat quality, insofar as they can be interpreted as evidence of lack of human disturbance of the environment. All of these data indicate that Mount Makiling is a valuable example of the biological diversity present on Luzon and that it is on a par with other global centers of biodiversity.

Attention has been recently given to the use of amphibians and reptiles as indicators of environmental health, as they are highly intolerant of severe changes in water quality (Alcala et al. 1995), which can be used in identifying key habitats for conserving biodiversity. One species, Woodworth's frog *(Rana woodworthi)*, classified as endemic and rare (Alcala 1986), was recorded at multiple study sites on Mount Makiling. Like most endemic Philippine forest frogs, the species is known to inhabit pristine and unpolluted creeks and rivers and easily succumbs to a deterioration in water quality due to pollution (pesticides, fertilizers, heavy metals, detergents, and other chemicals) or siltation resulting from soil erosion. Its continued presence on Mount Makiling therefore attests to the continued health of much of the reserve

Three skinks observed by project researchers on Mount Makiling are endemic and rare in the Philippines: the black-sided sphenomorphus *(Sphenomorphus decipiens)*, Steere's sphenomorphus *(S. steerei)*, and Cuming's eared skink *(Otosaurus cumingi)*. Four endemic species identified by the project represented new records on Mount Makiling: the white-spotted angle head *(Gonocephalus semperi)*, the smooth-skinned tree frog *(Philautus surdus)*, the Luzon forest frog *(Platymantis* sp. A), and the Mount Makiling forest frog *(Platymantis* sp. B). The project also recorded in the midmontane forests two new species that are rare, restricted range, single island endemics worthy of conservation (Alcala, personal communication 1996).

Only one threatened bird species, the Luzon tarictic hornbill *(Penelopides manillae)*, was observed at any of the project sites, although a number of near-threatened birds, such as the blue-breasted flycatcher *(Cyornis herioti)*, the blackish cuckoo-shrike *(Coracina coerulescens)*, the spotted wood kingfisher *(Actenoides lindsayi)*, the Philippine fairy bluebird *(Irena cyanogaster)*, the guaiabero fig-parrot *(Bolbopsittacus lunulatus)*, the spotted button-quail *(Turnix ocellata)*, the rufous coucal *(Centropus unifurus)*, and the Luzon bleeding-heart *(Gallicolumba luzonica)*, were observed (Collar and Andrew 1988). All of these species are Philippine endemics, five are strictly Luzon Island endemics, and eight are identified as "restricted range species."[2]

A number of factors affect bird diversity, including elevation, as earlier discussed. The tendency for diversity to vary inversely with altitude was clearly observed at the lower avian diversity in the mossy forest site near Mount Makiling's peak compared to higher diversity at the lowland sites, as in the dipterocarp forest on Dampalit. Whereas species richness varies with altitude, it is relatively uniform within habitat types, although a high level of species turnover is found in habitat ecozones (Navarro 1992). This was clearly seen in agroforestry areas on Mount Makiling, where high levels of observed diversity reflected the mixing of species from two different habitat types (forests and agroecosystems) in one ecozone. As these data indicate, avian diversity does not necessarily show a strong correlation with the lack of manmade disturbance. This is unlike the case with amphibians and reptiles, in which species diversity shows an inversely proportional relationship with human disturbance of the environment.

The conservation status of the mammals observed in our study suggests that the midmontane dipterocarp forest at 400 to 650 m asl is an important habitat because it harbors the Philippine pygmy fruit bat *(Haplonycteris fischeri)*, a species designated as vulnerable on the Red List of the World Conservation Union (also known as the IUCN; see IUCN 1994). Significantly, a number of endemic and rare species of horseshoe bats were also recorded in this zone. Other species with similar conservation values recorded in the study area were the lowland striped shrew-rat *(Chrotomys mindorensis)*, the southern Luzon giant cloud rat *(Phloeomys cumingi)*, the long-tailed macaque *(Macaca fascicularis)*, the Philippine brown deer *(Cervus mariannus)*, the Philippine warty pig *(Sus philippensis)*, and the large flying fox *(Pteropus vampyrus)*. More importantly, there were two new records *(Miniopterus schreibersii* and *Pipistrellus stenopterus)*. With bats, as with birds, an increase in elevation results in a decrease in diversity (but not necessarily endemicity). Among small mammals, however, an opposite association was observed wherein an increase in elevation results in an increase in diversity and endemicity.

Large and conspicuous mammals that require large tracts of forest habitat and for this reason are often utilized as mammalian indicators of biodiversity, are largely absent from the Philippines. The majority of the mammals found are bats and small mammals (such as rodents and shrews). However, the high diversity of bats and rodents in the Philippines may itself be a valuable indicator, as many are endemic forest species. These endemics are often restricted in distribution and are found mostly in primary or old growth forests. Some are highly restricted and are found only on specific mountains and islands. Others are highly specific to certain microhabitats, with peculiar trophic guilds feeding on worms, fruits, or leaves. A large number of these endemic species are found in highland forests. Moreover, small mammals and bats are sensitive to human disturbance of their habitats, leading to a decrease in endemicity and diversity (Rickart, Heaney, and Utzurrum 1991). Intolerant endemic small mammals in the forests may be replaced with introduced commensal species. This was clearly observed in our study sites on Mount Makiling, where the number of mammals decreased with increasing human disturbance.

Socioeconomic Determinants of Crop Biodiversity at the Farm Level

The social, economic, political, and biophysical environment determines a farmer's decision about what particular crop mix to plant on his farm. The social factors include personal and household attributes such as education, household size, the skills and training of the farmer, and farming experience. Economic considerations include labor availability, prices of both inputs and outputs, the presence of marketing facilities, and access to credit. Climatic factors, geographic accessibility, and soil suitability are among the biophysical considerations. The extent to which farm-level diversity is influenced by all of these factors was the concern of this part of the study.

This analysis was based on the use of data gathered in previous surveys of the MFR. Supplemental data were obtained from key informants, with some of the findings cross-checked with specific households in the area. A survey of key informants was carried out to assess public opinions on a number of biodiversity-related issues. Finally, the informants also were asked to assess the project scientists' identification of indicator species.

Previous surveys in the MFR showed that farming accounts for only 40 percent of the average forest resident's household income. The few fruit plantations in the area are mostly owned by nonresidents and managed by tenants. The dominant land use system in the forest reserve is agroforestry (Cruz, Francisco, and Torres 1991; Sumalde, Francisco, and Fermintoza 1992; Carandang and Lawas 1992), which refers to a combination of annuals (such as root crops and vegetables) and fruit trees (such as *lanzones*, citrus, rambutans, coconuts, and coffee). The average farm size is approximately two hectares. The primary mode of land acquisition in the MFR is through inheritance. Only 30 percent of the resident farmers claim to have acquired land by clearing the forest (Dizon 1992). But the establishment of perennial crops is seen by many (residents and policymakers alike) as a means of staking a claim in the reserve. Indeed, the local residents' claims over what they have planted are being used today as the main argument for their right to stay in the forest reserve. This means that financial reasons or conservation considerations may only be secondary factors in the decision to plant perennial crops in the agroforestry system.

Biodiversity and Socioeconomic Attributes

To analyze the relationship between biodiversity and socioeconomy, first the residents of Mount Makiling were differentiated using key socioeconomic variables, including farmer's age, farm size, number of years farming, household income, education, and household size.[3] A crop diversity index of qualitative variation was then calculated using respondent recall, which measured variations in numbers and kinds of crops. This index was then calculated for subsets of the resident population, broken down by each of the key socioeconomic variables. Finally, correlation between these variables and biodiversity was calculated using simple and multiple regression analyses. The succeeding discussion presents a simple test of the relationship between biodiversity as measured by CD-IQV and selected socioeconomic variables (see table 7.2).

What do the results of this analysis tell us? First, the relationship between household income and crop diversity is inverse. If we associate the higher income with a more market-oriented system of agriculture and the lower income with a more subsistence oriented system, the observed difference in crop diversity is what we usually expect to see in market-oriented versus subsistence-oriented agriculture.

Similarly, there is an inverse relationship between crop diversity and education. The more educated the farmer is, the greater is his tendency to specialize in fewer but more lucrative commercial crops. Higher education also allows the farmer to take more risks with fewer crops, as he has (due to his/her education) more potential nonfarm sources of income. Crop diversity also is inversely correlated with the age of the household head. The highest diversity index was noted among the younger households, whereas a relatively lower diversity index was observed among the older farmers. The relationship between age of household head and the crop biodiversity index is thus negative. This means that younger farmers tend to have more diverse farms than the older farmers do. This counterintuitive finding is explained by patterns of land distribution. Young farmers do not necessarily cultivate young or new farms. Division and devolution of farm holdings typically takes place when children marry. Newly married members of the family are often given an established farm that already has diversified plantings on it. In addition, they are likely to be given a small farm, and with a smaller farm, more intensive cultivation through diversified cropping systems is likely, leading to greater crop biodiversity in young households.

The relationship between farm size and crop diversity is less clear. Although a high CD-IQV diversity index is associated with the largest farm size, the difference in the diversity levels between small and medium farms is minute and insignificant. There is also no discernible relationship between years of farming and crop diversity levels. The highest diversity level was observed among the households that had been involved in farming for only an average amount of time, and low diversity was noted among farmers with both fewer and more years of experience.

For the regression analysis, three measures of crop diversity were used: CD-IQV; the Shannon-Weiner index, based on number of crop species (the crop diversity index, or CDI); and the simple number of plant species (plant diversity or PD). Each of the three measures was treated as the dependent variable in the farming household. The independent variables were the household's socioeconomic attributes: age, sex, years of farming, and education of the household head as well as household income, household size, and farm size. Several different regression models were tried: linear, log linear, linear log, and double log.[4] Of these models, the log-linear model has the best fit and the highest value of R^2. Further, it has the highest number of significant independent variables. Table 7.2 shows the results of the regression analyses. In all three regression runs, however, the explanatory variables included in the model account for a large proportion of the variation in the level of the dependent variable (crop diversity). This is indicated in the high values of the coefficient of determination (R^2). The similarity in the R^2 values of the CD-IQV and the CDI reflect the fact that both refer to economic crops and hence may be considered comparable.

The most significant explanatory variables in the regression analysis are farm size and household size, both of which have positive coefficients. An increase in household size means that a larger labor force is available to attend to crop diversification, whereas an

Table 7.2 *Regression analysis of factors affecting crop and plant diversity*

Variable	CD-IQV[a]	CDI[b]	PD[c] × (No. of plots)
Constant	3.897	1.057	4.174
Age	0.0100[d]	-0.0023	-0.0071
Sex	0.2813	-0.155	0.168
Education	-0.1110[e]	0.0047	-0.0019
Household income	-4.33E-07	-7.19E-06[e]	-2.27E-05[f]
Household size	0.0684[g]	0.0401[d]	0.1150[e]
Farm size	0.0341[e]	0.0095[g]	0.020[g]
Farming (yrs.)	-0.0076	0.0023	-0.0021
F-value	3.4[d]	2.74[g]	9.05[f]
R-square	82.95	79.32	92.69
Adj. R-square	59.07	50.36	82.45

[a]IQV = Index of qualitative variation using survey data (Torres 1992)
[b]CDI = Crop diversity index based on field data (Gruezo 1995)
[c]PD = Plant diversity based on number of plant species in the farm as identified by Gruezo (1995).
[d]Significant at 10% confidence level
[e]Significant at 5% confidence level
[f]Significant at 1% confidence level
[g]Significant at 15–20% confidence level

increase in farm size means a greater ability to accommodate an increase in the number of different crop types cultivated.

The CDI and PD are both negatively affected by household income. In both cases, the regression coefficient is quite small, however, which means that the extent of influence may not be substantial. The negative value of the coefficient reflects the fact that higher household income is often associated with monocropping or plantation-type agriculture. It has been noted on Mount Makiling that the high-income farmers tend to adopt plantation-type agriculture. This type of agriculture also requires more capital, which only households with higher incomes can manage. Further, risk increases with fewer crop species, and the lower-income groups apparently are not willing to take this much risk. As noted earlier, the lower the income the greater the tendency for farmers to diversify, which reduces their exposure to risk.

Age has a positive coefficient at a significance level of 15 to 20 percent, which means that 80 to 85 percent of the time an older farmer will have more crop diversity in his farm. This is not consistent with the results obtained earlier when considering only age in explaining the level of diversity. It appears that in the regression analysis the interaction of other socioeconomic factors has reversed the relationship between age and diversity.

In general, the results from the three types of regression analyses are not greatly different from one another. The variables that emerged from the analysis as significant determinants of crop diversity are consistent with what we expected. The findings lend support to what many observers have been thinking all along, namely that farm size, household size, household income, and to some extent age and education all are major determinants of crop diversity. These variables account for 80 to 90 percent of the observed variation in the value of the dependent variable, crop diversity.

Finally, the results of this analysis attest to the reliability of household survey data on diversity given the great similarity in the correlation coefficients obtained by the two crop diversity measures—the one based on household survey and the other based on field measurements. For a more thorough comparative evaluation of the two measures, however, additional factors must be brought into the picture, such as the cost of data collection and consistency in the reliability of the measures. Nevertheless, the preliminary findings do suggest that household-generated data may provide measures of biodiversity equally as useful as the more time-consuming plot inventories.

Summary and Conclusion

Changing Views of Biodiversity Conservation

The results of this study show the need for a more sophisticated contextualization of biodiversity measures in terms of the economic, ecological, and social functions of biodiversity itself. The Mount Makiling study indicates that diversity levels in less human affected ecosystems such as mossy forests and midmontane dipterocarp forests are approximately the same as those of the agroforestry system. These ecosystems, however, differ in their social, economic, and ecological conditions, with the diversity of the agroforestry system having strong economic and social functions. This comparative analysis of diversity raises important questions, including what constitutes "good" versus "bad" biodiversity and what kind of biodiversity should be maintained. These questions will only be answered if researchers and stakeholders alike become more knowledgeable about the various functions of biodiversity at all hierarchical levels (farm, ecosystem, and landscape).

The results of this study also indicate the need to combine traditional quantitative and statistical measures of biodiversity with newer, more rapid, qualitative methods. One way to develop rapid assessment methods is to use "surrogate measures" that are significantly correlated with diversity levels, such as indicator species. Selection criteria used in the identification of indicator species on Mount Makiling included: (1) the occurrence of the species over a wide altitudinal range or a variety of ecozones; (2) the opposite situation, in which the species has a very limited occurrence, has unique habitat requirements, or shares a similar morphological structure with other unrelated species within the ecozone; and (3) information from key informants on economic uses as well as spatial and temporal distributions of identified species; and (4) farmers' perceptions of what the indicator species' abundance or absence indicates. The biologists' list of indicator species was reviewed by key informants from the MFR. Information on plant parts used, products consumed and used, and the abundance of these species was also obtained and discussed. The key informants generally supported the selection of indicator species by the project scientists. With input from farmers and scientists alike, the final list of indicator species seemed to be quite useful in reflecting changes in the biotic and abiotic conditions as well as species and community composition of ecozones on Mount Makiling. The appendix (table 7.5A) presents the floral indicators of biodiversity found on Mount Makiling as perceived by both the farmers and the scientists. Although a more comprehensive look at the differences between scientists and farmers in perceiving biodiversity is needed, locally recognized indicator species still seem to offer an underutilized tool that can help identify valuable and declining habitats.

Finally, the results of this study demonstrate the importance of analyzing diversity based on more than statistical analysis of species counts alone. Other factors, such as endemicity, tolerance of disturbance, species-specific conditions, and community structure, also need to be examined. And whereas rapid assessments are useful, there is also a need to study Mount Makiling's diversity by conducting more comprehensive and long-term research.

Society and Environment

This study has demonstrated that the ecosystems surrounding Mount Makiling are rich in a diversity of rare and endemic species. At the same time, this region has been the home of agriculturalists for centuries. And, importantly, the complex agroforestry systems of the resident agriculturalists provide habitats for a wide range of species. In fact, species diversity in the anthropogenic agroforestry systems is higher than that found in the grasslands or mossy forests and is surpassed only by the dipterocarp forests. These findings suggest that perhaps the most important question to be explored in this region is how floral and faunal diversity has coevolved over time in association with anthropogenic changes in the environment. The answer to that question will require a long-term study of biodiversity change in the Mount Makiling area, a study that will be facilitated by the baseline data provided in this chapter.

Despite the need for long-term studies of ecological change, important short-term patterns that need emphasizing have emerged from this study. Perhaps one of the most significant findings is a positive correlation between an increase in farm and household size, on one hand, and biodiversity on the other. Conversely, a negative correlation is found between increased wealth and education compared to biodiversity. These findings suggest that farmers with more labor and land resources are able to diversify their crops, thereby increasing levels of biodiversity across the landscape. Conversely farmers with more education and greater financial resources tend to be more profit oriented and therefore are more inclined to take on the risks and promise of large rewards inherent in planting monocultures. Therefore, it may be necessary, as farmers become more educated, to develop policies and extension services that stress the trade-offs between landscape homogeneity and economic benefits. As the study of the upper Citarum river basin in Indonesia demonstrates (see Gunawan, Parikesit, and Abdoellah, this volume), short-term economic gains from intensive cash crop agriculture often have negative long-term effects on ecological stability.

Government Policy

These findings point toward the paradoxical consequences that conflicting state policies may bring about at the regional landscape level, as Gunawan Parikesit, and Abdoellah (this volume) found to be the case in Indonesia. In the Mount Makiling case, on the one hand the state threatens biodiversity through agricultural policies aimed at intensification and monocultivation for the market. But on the other hand the state admonishes local people for threatening biodiversity through their use of the land and resources within the forest reserve. Yet these same local subsistence uses of forest resources enable famers to be less dependent on external markets and thus less involved in market-oriented monocropping. Farmers who can rely on forest resources for subsistence tend to have much more diverse gardens than those who cannot. And the combination of subsistence use of forest resources and a highly diverse cultivation strategy produces an overall positive effect on floral and faunal diversity at the landscape level. This raises questions, in short, about the "accepted wisdom" that the best way to conserve forests is to delink local communities from them through increased market involvement. The results of both this study and the Indonesian study suggest that the current development and conservation paradigm, which emphasizes increased farmer involvement in the market through intensive monocultures and total exclusion of local people from state forests, may not facilitate the conservation of biodiversity.

Future policies in the Mount Makiling area would also do well to draw on the findings of Vergara (this volume) regarding the optimal shape and intensity of habitat fragmentation to promote biodiversity conservation in anthropogenic ecosystems. An innovative policy would include a landscape-wide vision encompassing the relationship between farm and forest. Such a reorientation of policy should emphasize optimizing habitat fragmentation through agriculture and at the same time support forest conservation through the design of buffer zones that allow human uses of forest resources. Ultimately, the links between the forest and the farm would contribute to more complex and diverse agroecosystems and ideally more complex diversity at the landscape level.

Notes

1 There often appears to be an inverse relationship between diversity and frequency of species. Whereas the agroforestry areas have moderate diversity but high frequency of species, the midmontane dipterocarp forests have high diversity but low frequency per species. The *Imperata* grasslands have low diversity but high frequency per species.

2 The high number of threatened or endangered endemics in the study area suggests that Mount Makiling should be considerated for official classification as an important bird area (IBA) in the Philippines.

3 This analysis was based on data from Torres 1992.

4 The log-linear and the linear-log forms refer to a functional relationship wherein the dependent and independent variables, respectively, are transformed into logarithmic functions.

CHAPTER SEVEN APPENDIX

Table 7.1A *Amphibians recorded in the Mount Makiling Forest Reserve*

Common Name	Family	Species Name	Remarks
Philippine woodland frog	Ranidae	*Rana magna macrocephala*	Endemic
Woodworth's frog	Ranidae	*Rana woodworthi*	Endemic
Variable-backed frog	Ranidae	*Rana sigmata similis*	Endemic
Common small-headed frog	Ranidae	*Occidozyga laevis*	Commensal
Common forest frog	Ranidae	*Platymantis dorsalis*	Endemic
Corrugated forest frog	Ranidae	*Platymantis corrugatus*	Endemic
Luzon forest frog	Ranidae	*Platymantis sp. A*	Endemic
Mt. Makiling forest frog	Ranidae	*Platymantis sp. B*	Endemic
Common tree frog	Rhacophoridae	*Polypedates leucomystax*	Commensal
Smooth-skinned tree frog	Rhacophoridae	*Philautus surdus*	Endemic
Luzon narrow-mouthed frog	Microhylidae	*Kaloula kalingensis*	Endemic

Total species = 12

Total endemic species = 10

Table 7.2A *Reptiles recorded in the Mount Makiling Forest Reserve*

Common Name	Family	Species Name	Remarks
Philippine bent-toed gecko	Gekkonidae	*Cyrtodactylus philippinicus*	Endemic
Jagor's sphenomorphus	Scincidae	*Sphenomorphus jagori*	Endemic
Black-sided sphenomorphus	Scincidae	*Sphenomorphus decipiens*	Endemic
Steere's sphenomorphus	Scincidae	*Sphenomorphus steerei*	Endemic
Cox's sphenomorphus	Scincidae	*Sphenomorphus coxi*	Endemic
Cuming's eared skink	Scincidae	*Otosaurus cumingi*	Endemic
Spotted green tree skink	Scincidae	*Lamprolepsis smaragdina*	Resident
Common mabouya	Scincidae	*Mabuya multifasciata*	Commensal
Spiny waterside skink	Scincidae	*Tropidophorus grayi*	Resident
White-spotted angle head	Agamidae	*Gonocephalus semperi*	Endemic
Malay monitor lizard	Varanidae	*Varanus salvator*	Resident
Spot-bellied short-headed snake	Colubridae	*Oligodon modesdum*	Endemic
Paradise tree snake	Colubridae	*Chrysopelea paradisii*	Resident
Elongate-headed tree snake	Colubridae	*Ahaetulla prasina*	Resident
Common rat snake	Colubridae	*Elaphe erythrura*	Resident
Northern triangle-spotted snake	Colubridae	*Cyclocorus lineatus*	Endemic
Philippine cobra	Elapidae	*Naja philippensis*	Endemic
Barred coral snake	Elapidae	*Calliophis calligaster*	Endemic
Philippine pit-viper	Viperidae	*Trimeresurus flavomaculatus*	Endemic

Total species = 19

Total endemic species = 12

Table 7.3A *Birds recorded in the Mount Makiling Forest Reserve*

Common Name	Local Name	Family	Species Name	Remarks
Osprey	Lawin	Pandionidae	*Pandion haliaetus*	Migrant
Besra sparrowhawk	Lawin	Accipitridae	*Accipiter virgatus*	Resident
Philippine serpent eagle	Agila	Accipitridae	*Spilornis holosphilus*	Endemic
Philippine falconet	Ibong-bingi	Falconidae	*Microhierax erythogenys*	Endemic
Red junglefowl	Labuyo	Phasianidae	*Gallus gallus*	Resident
Blue-breasted quail	Pugo	Turnicidae	*Turnix suscitator*	Resident
Barred button-quail	Pugo	Turnicidae	*Turnix ocellata*	Luzon-endemic
Pompadour green pigeon	Punay	Columbidae	*Treron pompadora*	Resident
Green imperial pigeon	Balud	Columbidae	*Ducula aenea*	Resident
Philippine cuckoo-dove	—	Columbidae	*Macropygia tenuirostris*	Endemic
Common emerald dove	Umamban	Columbidae	*Chalcophaps indica*	Resident
Luzon bleeding heart pigeon	Lagaran	Columbidae	*Gallicolumba luzonica*	Luzon-endemic
White-eared brown dove	Limukon	Columbidae	*Phapitreron leucotis*	Endemic
Amethyst brown dove	Kukuk	Columbidae	*Phapitreron amethystina*	Endemic
Cream-bellied fruit dove	Punay	Columbidae	*Ptilinopus occipitalis*	Endemic
Turtle dove	Batu-bato	Columbidae	*Streptopelia sp.*	Resident
Zebra dove	Batu-bato	Columbidae	*Geopelia striata*	Introduced?
Guaiabero	Bubutok	Psittacidae	*Bolbopsittacus lunulatus*	Endemic
Philippine hanging parrot	Kolasisi	Psittacidae	*Loriculus philippensis*	Endemic
Hodgson's hawk-cuckoo	—	Cuculidae	*Cuculus fugax*	Resident
Philippine brush cuckoo	—	Cuculidae	*Cacomantis sepulcralis*	Endemic
Plaintive cuckoo	—	Cuculidae	*Cacomantis merulinus*	Resident
Drongo cuckoo	—	Cuculidae	*Surniculus lugubris*	Resident
Lesser coucal	Sabukot	Cuculidae	*Centropus benghalensis*	Resident
Philippine coucal	Sabukot	Cuculidae	*Centropus viridis*	Endemic

(continued)

Table 7.3A, *continued*

Common Name	Local Name	Family	Species Name	Remarks
Rufous coucal	—	Cuculidae	*Centropus unirufus*	Luzon-endemic
Scale-feathered malkoha	Manuk-manuk	Cuculidae	*Phaenicophaeus cumingi*	Luzon-endemic
Rough-crested malkoha	Manuk-manok	Cuculidae	*P. superciliosus*	Luzon-endemic
Philippine scops owl	Kuwago	Strigidae	*Otus megalotis*	Endemic
Philippine hawk owl	Kurakpaw	Strigidae	*Ninox philippensis*	Endemic
Australasian grass owl	Kuwago	Tytonidae	*Tyto longimembris*	Resident
Great eared nightjar	—	Caprimulgidae	*Eurostopodus macrotis*	Resident
Savanna nightjar	—	Caprimulgidae	*Caprimulgus affinis*	Resident
Philippine frogmouth	Tiyanak?	Podargidae	*Batrachostomus septimus*	Endemic
Purple needletail	Layang-layang	Apodidae	*Hirundapus celebensis*	Resident
Fork-tailed swift	Layang-layang	Apodidae	*Apus pacificus*	Resident
Island swiftlet	Balinsasayao	Apodidae	*Collocalia vanikorensis*	Resident
Glossy swiftlet	Balinsasayao	Apodidae	*Colocallia esculenta*	Resident
Pygmy swiftlet	Layang-layang	Apodidae	*Collocalia trogodytes*	Endemic
Philippine swiftlet	Layang-layang	Apodidae	*Collocalia mearnsi*	Endemic
Philippine trogon	—	Trogonidae	*Harpactes ardens*	Endemic
Spotted wood kingfisher	—	Alcedinidae	*Actenoides linsayi*	Endemic
White-throated kingfisher	Salak-sak	Alcedinidae	*Halcyon smyrnensis*	Resident
Luzon tarictic hornbill	Tariktik	Bucerotidae	*Penelopides manillae*	Luzon-endemic
Chestnut-headed bee-eater	—	Meropidae	*Merops viridis*	Resident
Blue-tailed bee-eater	—	Meropidae	*Merops philippinus*	Resident
Coppersmith barbet	—	Capitonidae	*Megalaima haemacephala*	Resident
Philippine pygmy woodpecker	Manuktok	Picidae	*Dendrocopos maculatus*	Endemic
Crimson-backed woodpecker	Manuktok	Picidae	*Chrysocopaltes lucidus*	Resident
Sooty woodpecker	Manuktok	Picidae	*Mulleripicus funebris*	Endemic

(continued)

Table 7.3A, *continued*

Common Name	Local Name	Family	Species Name	Remarks
Red-breasted pitta	—	Pittidae	*Pitta erythrogaster*	Resident
Barn swallow	Layang-layang	Hirundinidae	*Hirundo rustica*	Migrant
Barred cuckoo shrike	Uwak-uwakan	Campephagidae	*Coracina strita*	Resident
Blackish cuckoo shrike	Uwak-uwakan	Campephagidae	*Coracina coerulescens*	Endemic
Pied triller	Ibong-pari	Campephagidae	*Lalage nigra*	Resident
Philippine fairy bluebird	—	Irenidae	*Irena cyanogaster*	Endemic
Yellow-wattled bulbul	—	Pycnonotidae	*Pycnonotus urosticus*	Endemic
Yellow-vented bulbul	Pulangga	Pycnonitidae	*Pycnonotus goiavier*	Resident
Philippine bulbul	Tururiyak	Pycnonitidae	*Hypsipetes philippinus*	Endemic
Balicassiao	Balikasyaw	Dicruridae	*Dicrurus balicassius*	Endemic
Elegant titmouse	—	Paridae	*Parus elegans*	Endemic
Velvet-fronted nuthatch	—	Sittidae	*Sitta frontalis*	Resident
Stripe-headed rhabdornis	—	Rhabdornithidae	*Rhabdornis mysticalis*	Endemic
Black-crown tree babbler	—	Timaliidae	*Stachyris nigrocapitata*	Endemic
White-browed shortwing	—	Turdidae	*Barchyteryx montana*	Resident
Magpie robin	—	Turdidae	*Copsychus saularis*	Resident
White-browed shama	—	Turdidae	*Copsychus luzoniensis*	Endemic
Island thrush	—	Turdidae	*Turdus poliocephalus*	Resident
Pied chat	—	Turdidae	*Saxicola caprata*	Resident
Arctic warbler	—	Sylviidae	*Phylloscopus borealis*	Migrant
Philippine leaf-warbler	—	Sylviidae	*Phylloscopus olivaceus*	Endemic
Duboi's leaf-warbler	—	Sylviidae	*Phylloscopus cebuensis*	Endemic
Mountain leaf-warbler	—	Sylviidae	*Phylloscopus trivirgatus*	Resident
Oriental leaf-warbler	—	Sylviidae	*Acrocephalus orientalis*	Migrant
Striated grasshopper warbler	—	Sylviidae	*Locustella fasciolata*	Migrant

(continued)

Table 7.3A, *continued*

Common Name	Local Name	Family	Species Name	Remarks
Philippine tailorbird	—	Sylviidae	*Orthotomus castaneiceps*	Endemic
Grey-backed tailorbird	—	Sylviidae	*Orthotomus derbianus*	Endemic
Mountain tailorbird	—	Sylviidae	*Orthotomus cucullatus*	Resident
Tawny grassbird	Tinteryok	Sylviidae	*Megalurus timorensis*	Resident
Striated grassbird	Tinteryok	Sylviidae	*Megalurus palustris*	Resident
Bright-capped fantail-warbler	—	Sylviidae	*Cisticola exilis*	Resident
Oriental bush warbler	—	Sylviidae	*Cettia diphone*	Migrant
Blue-breasted flycatcher	—	Muscicapidae	*Cyornis herioti*	Luzon-endemic
Grey-streaked flycatcher	—	Muscicapidae	*Muscicapa griseisticta*	Migrant
Blue-baled blue monarch	Chiki-chiki	Monarchidae	*Hypothymis azurea*	?
Blue-headed fantail	—	Rhipiduridae	*Rhipidura cyaniceps*	Endemic
Yellow-bellied whistler	—	Pachycephalidae	*Pachycephala philippensis*	Endemic
Olive tree pipit	—	Motacillidae	*Anthus hodgsoni*	Migrant
Petchora's pipit	—	Motacillidae	*Anthus gustavi*	Migrant
Brown shrike	Tarat	Laniidae	*Lanius cristatus*	Migrant
Long-tailed shrike	Berdugo	Laniidae	*Lanius schach*	Resident
White-breasted wood-swallow	—	Artamidae	*Artamus leucorhynchus*	Resident
Coleto	Kuhling	Sturnidae	*Sarcops calvus*	Endemic
Olive-backed sunbird	Pipit-puso	Nectariniidae	*Nectarinia jugularis*	Resident
Purple-throated sunbird	Pipit-puso	Nectariniidae	*Nectarinia sperata*	Resident
Lovely sunbird	—	Nectariniidae	*Aethopyga shelleyi*	Endemic
Flaming sunbird	—	Nectariniidae	*Aethopyga flagrans*	Endemic
Buzzing flowerpecker	—	Dicaeidae	*Dicaeum hypoleucum*	Endemic
Pygmy flowerpecker	—	Dicaeidae	*Dicaeum pygmaeum*	Endemic
Bicolored flowerpecker	—	Dicaeidae	*Dicaeum bicolor*	Endemic

(continued)

Table 7.3A, *continued*

Common Name	Local Name	Family	Species Name	Remarks
Orange-breasted flowerpecker	—	Dicaeidae	*Dicaeum trigonostigma*	Endemic
Philippine flowerpecker	—	Dicaeidae	*Dicaeum australe*	Endemic
Striped flowerpecker	—	Dicaeidae	*Dicaeum aeruginosum*	Endemic
Eurasian tree sparrow	Mayang-bato	Passeridae	*Passer montanus*	Introduced
Mountain white-eye	Matang-isda	Zosteropidae	*Zosterops montanus*	Resident
Golden-yellow white-eye	Matang-isda	Zosteropidae	*Zosterops nigrorum*	Endemic
Philippine white-eye	Matang-isda	Zosteropidae	*Zosterops meyeni*	Endemic
Chestnut munia	Mayang-pula	Estrildidae	*Lonchura malacca*	Resident
White-breasted munia	Maya	Estrildidae	*Lonchura leucogastra*	Resident

Note: Netting results determine the number of individuals captured per net-night or net-day, wherein one net-night or net-day is equivalent to one net-set for one night or day.

Table 7.4A *Mammals recorded in the Mount Makiling Forest Reserve, 1993–96*

Common Name	Family	Species Name	Remarks
Common short-nosed fruit bat	Pteropodidae	*Cynopterus brachyotis*	Resident
Philippine pygmy fruit bat	Pteropodidae	*Halponycteris fischeri*	Endemic
Dagger-toothed flower bat	Pteropodidae	*Macroglossus minimus*	Resident
Musky fruit bat	Pteropodidae	*Ptenochirus jagorii*	Endemic
Common rousette	Pteropodidae	*Rousettus amplexicaudatus*	Resident
Large flying fox	Pteropodidae	*Pteropus vampyrus*	Resident
Diademe roundleaf bat	Rhinolophidae	*Hipposideros diadema*	Resident
Philippine forest roundleaf bat	Rhinolophidae	*Hipposideros obscurus*	Endemic
Arcuate horseshoe bat	Rhinolophidae	*Rhinolophus arcuatus*	Resident
Big-eared horseshoe bat	Rhinolophidae	*Rhinolophus macrotis*	Resident
Small rufous horseshoe bat	Rhinolophidae	*Rhinolopus subrufus*	Endemic
Yellow-faced horseshoe bat	Rhinolophidae	*Rhinolophus virgo*	Endemic
Common bent-winged bat	Vespertilionidae	*Miniopterus schreibersii*	Resident
Common asiatic myotis	Vespertilionidae	*Myotis horsefieldii*	Resident
Orange-fingered myotis	Vespertilionidae	*Myotis rufopictus*	Endemic
Javan pipistrelle	Vespertilionidae	*Pipistrellus javanicus*	Resident
Narrow-winged pipistrelle	Vespertilionidae	*Pipistrellus stenopterus*	Resident
Luzon forest shrew	Soricidae	*Crocidura grayi*	Endemic
Asian house shrew	Soricidae	*Suncus murinus*	Commensal
Asian house mouse	Muridae	*Mus musculus*	Commensal

(continued)

Table 7.4A, *continued*

Common Name	Family	Species Name	Remarks
Common Philippine forest rat	Muridae	*Rattus everetti*	Endemic
Polynesia rat	Muridae	*Rattus exulans*	Commensal
Oriental house rat	Muridae	*Rattus tanezumi*	Commensal
Lowland striped shrew-rat	Muridae	*Chrotomys mindorensis*	Endemic
Southern Luzon giant cloud rat	Muridae	*Phloeomys cumingi*	Endemic
Long-tailed macaque	Cercopithecidae	*Macaca fascicularis*	Resident
Common palm civet	Viverridae	*Paradoxurus hermaphroditus*	Resident
Malay civet	Viverridae	*Viverra tangalunga*	Resident
Philippine warty pig	Suidae	*Sus philippensis*	Endemic
Philippine brown deer	Cervidae	*Cervus mariannus*	Endemic

Total species = 30

Total Philippine endemic = 12

Table 7.5A *Indicator floral species in the Mount Makiling Forest Reserve*

Indicator Species	Scientists' Perception	Farmers' Perception
Nettle plant (*Dendrochnide meyeniana*)	The overall appearance and abundance of this plant correlates, though in varying ways, with the extent and degree of human intrusion or disturbance across the Makiling landscape except in the grassland zone.	The plant is found all over the forest reserve but thrives best near bodies of water and in relatively cool environments. The plant is perceived to be not useful at all. It can cause contact dermatitis. Some claim that it is often cut because of this characteristic, while others claim they are not touched for the same reason. Its absence, therefore, can mean both human intrusion (if cutting is practiced) and human absence (if no cutting is practiced).
Sugar palm (*Arenga pinnata*)	The presence of this plant, especially in the dipterocarp midmontane and agroforestry zones, is correlated with the population density of large frugivorous animals such as the palm civet (*Paradoxurus hermaphroditus* subsp. *Philippinensis*). An abundance of fruit-bearing plants indicates less human intrusion in the ecozone, while lack of fruiting plants and the presence of immature plants indicate frequent collection of leaf midribs for broommaking by residents.	This species is abundant all over the forest reserve but is most abundant in the innermost part of the forest. The products derived from plant parts include brooms from leaf midribs, rope from leaf sheath fibers, vinegar from the inflorescence sap, an edible vegetable from the palm heart, candied sweets from fruits, and lumber from the tree trunk. This species is observed to grow abundantly from the seeds dispersed by the palm civet and is not considered scarce. As such, this may not be a good indicator of human disturbance.
Succulent forest floor herbs	The abundance and closely spaced growth formations of these succulent forest floor herbs (e.g. *Elastostema viridescens*) directly indicate no disturbance.	
Bird's nest ferns	The abundance and maturity of these plants (e.g., *Asplenium nidus*, *A. affine*, *A. cuneatum*, *A. excisum*, *A. pellucidum*, *A. polyodon*, *A. tenerum*, and *A. vittaeforme*) directly correlate with the absence of human disturbance, particularly in the mossy forest, where these plants are easily collected due to the short stature of their substrate trees.	Some respondents claimed that collection of this species is prohibited. Others say, however, that the species is often collected and as a result its population has decreased. This species is usually found in the dipterocarp and mossy forest zone, particularly clinging to big trees. The fern is collected for ornamental and medicinal purposes(e.g., *Asplenium nidus*).

(continued)

Table 7.5A, *continued*

Indicator Species	Scientists' Perception	Farmers' Perception
Ground orchids	The presence of an abundant population of these plants (e.g., *Habenaria malinta, H. hystrix, Calanthe lyroglossa, C. triplicata, Geodorum nutans,* and *Corymborchis confusa*), particularly in the dipterocarp midmontane forest, mossy forest, and perhaps in the grasslands, signifies less human intrusion.	Although the residents of the forest reserve point out that it is illegal to collect these species, (e.g., *Habenaria malinta*) they are often collected and sold as highly priced ornamentals. This activity has decreased their populations.
Vines and woody climbers	The abundance and maturity of vines and woody climbers signify an intact multilevel forest canopy and higher biodiversity (e.g., *Dioscorea alata, D. elmeri, Connarus semidecandrus, Archangelisia flava, Anamira cocculus, Cissus* sp., and *Strongylodon macrobotrys*).	Some species are collected for cordage and other uses such as shampoo (i.e., *Entada phaseoloides*), locally known as *gugu*. The decline in their population density can therefore indicate human intrusion and decrease.
Parasitic/strangler plants	Considered to be keystone species, these plant play a major role in maintaining ecosystem structure and integrity. The many species of *Ficus* (stranglers or not) and mistletoes provide a year-round supply of fruits to birds and mammals. In return, these predators ensure the perpetuation and maintenance of figs and mistletoe species in the exposed portions of forest canopy. The presence of rare parasitic plants such as *Rafflesia* indicates a low level of disturbance because even a slight perturbation in forest dynamics results in the loss of their hosts. The high host specificity of these parasitic plants makes them an even more sensitive indicator because their viney hosts depend on the presence of many large forest trees for their own proper growth and existence.	A number of informants have indicated that these species are left undisturbed as their wood quality is very low and they do not have any other economic value. These species are found all over the reserve, and their abundance or absence is thus not a good indicator of human intrusion.
Edible cane and ornamental rattans	As with the woody vines and lianas, a low density of rattan suggests a high level of human intrusion. These species have high economic value (e.g., *Calamus merilli, C. blancoi, C. dimorphocanthus, C. discolor, C. ornatus, C. usitatus, Daemonorops mollis,* and *D. ochrolepis*). Rattan is preferred to woody vines as a raw material for basketry, furniture, and other home products.	Although the collection of rattan is illegal, these plants are commonly collected for use in cottage industries (e.g., basketry and cane furniture making), as ornamental plants (e.g., *Calamus biscolor*), or for food (e.g., *Calamus merilli*). Rattan species are still abundant in the less accessible forest interior. The intervening factor of accessibility makes it difficult to associate their abundance or absence with the level of human disturbance.

8

Use of Global Legal Mechanisms to Conserve Local Biogenetic Resources: Problems and Prospects

MICHAEL R. DOVE

THE ASSOCIATION OF BIODIVERSITY conservation with indigenous peoples at the so-called periphery of the global system is well illustrated by the Melaban Kantu', a tribal people of West Kalimantan, Indonesia. They cultivate not irrigated rice fields, which have been the subject of extensive government interventions throughout Asia focusing on the introduction of high-yielding varieties of rice, but dryland (and some swampland) swiddens, which have received virtually no extension inputs, and which contain exclusively traditional upland rice varieties. They still retain many so-called animistic beliefs regarding the spirit of the rice, which—along with a subsistence pattern that emphasizes diversification—contributes to their planting more than one dozen different varieties of rice per household. Each household also plants an average of nearly two dozen different nonrice "relishes." This emphasis on diversification also is reflected in the fact that each household makes two to three separate swiddens each year, each located in a different microenvironment.

An earlier version of this chapter first appeared as M.R. Dove, Center, periphery and biodiversity: A paradox of governance and a developmental challenge. In *Intellectual Property Rights and Indigenous Knowledge*, edited by S.B. Brush and D. Stabinsky, 41–67 (Washington D.C.: Island Press, 1996).

In addition to their swiddens, the Kantu' cultivate Para rubber *(Hevea brasiliensis)*: most households own four to five separate rubber gardens, each averaging less than one hectare and containing several hundred trees. The rubber trees are unselected varieties, descended directly, in many cases through self-propagation, from seeds imported from Brazil at the end of the nineteenth century. The rubber gardens are not clean weeded, and they contain, in addition to the rubber trees, a wide variety of noncultigens, many of economic use, such as edible bamboos and ferns. The productive mixture of plants in the gardens is an attraction to wild animals, and they are a prime hunting ground for the bearded pig *(Sus barbatus)*, the most prized game animal of the Kantu'.

Swiddens and rubber gardens aside, the territory of the Kantu' is comprised mostly of secondary forest but includes pockets and strips of primary forest. The reasons for protecting these remnants are variously economic (where valuable fruit, timber, or *illipe* nut [*Shorea* spp.] trees are present), ecological (narrow strips of forest are usually left standing along the banks of streams and rivers to protect the waterways from erosion and blockage), and ritual (e.g., burial grounds and areas known to be the abode of *antu*, "spirits," always are protected). A variety of plants in these pockets and strips are exploited for food, craft, medicinal, and ritual purposes. The Kantu' traditionally communicated with their principal deities through the medium of seven species of forest birds, all of which—along with a number of other forest fauna—they are ritually forbidden to kill.

These are not atypical of the circumstances of biodiversity conservation by indigenous peoples in the Third World,[1] circumstances that are the current focus of attention by scholars and activists in the international development and scientific community (Zerner 2000). There is anxiety in this community over the current nature of exploitation of biogenetic resources by indigenous communities in less developed countries and by business and academic communities in the more developed ones (Laird 2002). It is suggested that exploitation by the former is leading to degradation and that this is exacerbated by the latter's failure to recognize and compensate these communities for their stewardship of these resources. It is further suggested that if due recognition and compensation were provided to these indigenous communities their purported degradation of these resources would diminish. One mechanism that has been proposed for this compensation and recognition is intellectual property rights (IPRs) (Greaves 1994).[2]

The proposal to use intellectual property rights to address the issues of biodiversity maintenance and socioeconomic equity in the less developed countries is sound, assuming that its premises are correct. But are they, particularly as regards the responsibility that they place on the indigenous community? In 1993, I met with representatives of a tribal-based nongovernmental organization (NGO) in Kalimantan, Indonesia, regarding a land conflict in the subdistrict from which most of them hailed. The NGO members reported that six different timber concessionaires had formally laid claim to 100 percent of the land of this subdistrict, including the land under their swiddens, rubber gardens, fruit groves, and even the villages themselves. In a spirit of compromise, the tribesmen offered the concessionaires free access to all remaining primary forest in their subdistrict as long as the lands currently in agricultural use were left in their hands, but the concessionaires refused. In response, the local tribesmen began felling the primary forest themselves and planting it with rubber to strengthen their future claims against the concessionaires.

The timber concessions were all held by Java-based corporations, and it is to Java that all income flowed. The central, national government justified the granting of such concessions in terms of the purported beneficial impact on the local economy, but a representative of the local, provincial planning office Direkori Badan Perencanaan Pembangunan Daerah (the Directory of Development Planning Board, or BAPPEDA) told me that ten years' experience with the logging concessions had demonstrated that they were totally without benefit to the local communities.

These logging concessions clearly had an enormous impact on biodiversity conservation and the rights of local communities. Equally clearly, intellectual property rights were not the appropriate mechanism with which to rectify this situation; they would simply have been beside the point. Nor is the case of the Kalimantan logging concessions unique; indeed, I suggest that this case is more the rule than the exception in degradation of biodiversity and community rights in less developed countries. This raises some serious questions about the wisdom of proposals to use intellectual property rights to address this degradation, and it suggests that its premises need to be reexamined.

Such a reexamination is the purpose of this analysis. I focus on a number of key questions. In particular, what is the role of marginality in biodiversity and what, thus, is the impact on biodiversity of the incorporation of marginal communities into broader political-economic systems? Can the same system that threatens biodiversity and indigenous rights also save them? Who supports the proposed use of intellectual property rights and why? And do resources flow into marginal communities as easily as they flow out or them?

My thesis is that the use of intellectual property rights to address the issues of degradation and inequity faces two serious obstacles: first, compensation for them is likely to fail to reach the indigenous peoples responsible for biodiversity conservation; and second, if compensation does reach them, it is likely to change them and undermine the basis for biodiversity conservation in the process. I argue that a possible solution to this dilemma lies in problematizing the act of intervention and the relationship between elites and indigenous communities in the less developed countries.

I begin with an outline of some of the implicit premises of the proposed use of intellectual property rights. I then discuss the impact on community resource rights and biodiversity conservation of, first, a more attenuated relationship with the world system and, second, a more intense relationship with the world system. Next I attempt to reframe the question of biodiversity conservation and community rights and conclude by suggesting some alternative means for coping with it.

The Premises of the Use of Intellectual Property Rights

The proposed use of intellectual property rights to address issues of biodiversity conservation and community rights in the developing world is based on one or more of a number of complex premises. To begin with, a key premise is that the industrialized countries of the world are unfairly exploiting the biological and intellectual resources in the less developed countries and that the anthropologists and other international scientists who work in these communities are major parties to this exploitation (e.g., Posey 1990). It is argued that the resources were typically appropriated without compensation in the first place, that valuable products have been developed from them, and that in many cases these manufactured products are being sold at high prices back to the communities that developed the original biogenetic resource.

A second premise is that this pattern of resource appropriation by the industrialized nations has contributed to the impoverishment of these indigenous communities and that this in turn has led to the overexploitation and thus degradation of the resources (Nations 1988; Blum 1993). A corollary assumption is that there are no other causes of degradation, that they are all confined to the local community. This is reflected in the wide support given to proposals for in situ conservation (Altieri and Merrick 1987; Brush 1991, 1992b; Oldfield and Alcorn 1989), the very definition of which implies that it is *not* necessary to look for solutions beyond the individual farm or community.

A third premise is that the community's poverty can be alleviated and resource degradation diminished or reversed through payment of compensation by the industrialized nations based on the Western concept of intellectual property rights. A corollary assumption is that payment of this compensation is not problematic and that resource flows into these communities are as efficient as resource flows from them.[3]

A fourth premise is that it is possible to make this compensation without adversely affecting those distinctive aspects of the community that contribute to the creation and maintenance of biodiversity. A corollary assumption is that there is no necessary relationship between the conservationist character of these community structures and the attenuated character of their involvement with national and international political-economic structures.

A fifth and final premise is that if this compensation is made to the national government as opposed to the local community directly, it will still have the same beneficial effect on biodiversity conservation by local communities. In other words, it is assumed that the state and the people are one, at least with respect to this issue.

Table 8.1 *The premises behind the proposed extension of IPRs to indigenous communities*

Premises		Critique
The MDCs are unfairly exploiting the biogenetic resources of indigenous communities in the LDCs, and anthropologists and ethnobotanists are major parties to this exploitation.	1	This greatly exaggerates exploitation by MDC scientists relative to exploitation by LDC political-economic elites.
This exploitation by the MDCs impoverishes indigenous communities, and they in turn (perforce) degrade their local resource-base.	2a	Resource degradation is due more to commercial extraction by national and international political-economic elites.
The immediate cause of this degradation is the local community.	2b	This degradation impoverishes indigenous communities, not the reverse.
This poverty can be alleviated and the degradation diminished through the MDCs' payment of compensation, based on the concept of intellectual property rights.	3a	Such compensation does not address the fact that most impoverishment and degradation is caused by political-economic elites.
Resource flows into these communities are as efficient as resource flows out of them.	3b	There are structural impediments to the flow of resources into such communities.
Payment of compensation will not threaten the indigenous structures associated with biodiversity conservation.	4a	Any payment of compensation necessarily entails some alteration of center-periphery relations.
There is no relationship between the conservationist character of these communities and the attenuated character of their involvement with national and international political-economic structures.	4b	Any such alteration is potentially problematic for biodiversity conservation by peripheral communities because this is functionally related to location at the periphery.
Compensation can be made to the national government as opposed to the indigenous community and it will still have the same beneficial effect on biodiversity conservation.	5	It is likely that indigenous communities and national governments will not share identity of purpose with respect to biodiversity conservation, so payment of compensation to the latter will not achieve the same goals as will payment to the former.

I suggest that all of these premises are partially or wholly untenable (table 8.1) and that this throws into question the use of intellectual property rights to the end proposed. One of the main problems with this proposed use is a misunderstanding of the implications of greater versus lesser incorporation of indigenous communities in the less developed countries into broader political-economic systems.

Nonincorporation

A key premise of the proposed use of intellectual property rights to address issues of biodiversity conservation and the rights of indigenous communities is that the extension of these rights to indigenous peoples in less developed countries is *not* problematic. The implication is that the global system that proposes to extend these rights, and the indigenous communities that are the intended beneficiaries, are structurally similar members of the same integrated system. I suggest, rather, that the global system and these indigenous communities are structurally *dissimilar* members of a more loosely articulated system. Their respective places within this system have been termed "center" and "periphery." Although the rationale behind the use of intellectual property rights pays no attention to the distinction between center and periphery, I suggest that it is real and determinant. Inattention to this distinction is a function of a paradoxical tendency among scholars and planners to insist that systems either are all embracing (e.g., the global system) or unconnected (e.g., indigenous communities). The concept of a differentiated system, with relations obtaining among dissimilar members, is relatively undeveloped in the international scientific and development community. An exception is in anthropology, especially the anthropology of Southeast Asia.

Anthropologists long ago demonstrated the dependence of centralized Southeast Asian states upon relations with peripheral communities such as hill tribes.[4] These peripheral groups furnished lowland centers with wives, warriors and enemies, trade goods and plunder, and shamanism and sorcery. They furnished developing states with an "other" that helped the state to establish its identity, its existence. The same kind of interdependence between core and hinterland, or center and periphery, characterizes the contemporary relationship between the global political-economic system and its periphery with respect to biodiversity. The marginality of the periphery is structural, it is integral to its relations with the rest of the global system, and it is associated with biodiversity. Brush (1992b, 1624) notes that "In many areas, the farmers who nurture crop genetic resources are economically and technologically isolated, and often they are members of outcast ethnic minorities." Elsewhere Brush (1993, 664) observes "Groups that control the greatest wealth of indigenous knowledge and biological resources may be the most marginalized by nation states, e.g. Kurds of Iraq or Mayan peasants of Guatemala."

The concentration of biodiversity at the periphery of the global system is not accidental. Semiautonomy and marginality from the global system both permit and are permitted by biodiversity. This is exemplified by high-diversity systems of cultivation such as swidden agriculture and by the combination of subsistence-oriented swiddens with market-oriented cash cropping in even more complex and diversified "composite" systems of agriculture, such as described earlier for the Kantu' of Kalimantan, Indonesia. The high index of agricultural biodiversity at global peripheries is necessitated by lack of access to the safety nets of national and international centers, which obliges the peripheries to pursue risk-adverse and thus diversified agricultural strategies. This strategy is permitted by the peripheries' relative freedom from the constraints of global markets, which demand genetic homogeneity in order to reap economies of scale. As Norgaard (1987, 103; 1989, 206) suggests, incorporation into the global system characteristically sacrifices diversity and stability at the microlevel to achieve diversity and stability at the macrolevel. Paradoxically, however, this macrolevel diversity and stability depends on continued access to—and thus maintenance of—biodiversity at the local level. The importance of this access is reflected in the efforts—the focus of the current study—to compensate indigenous communities for knowledge and maintenance of biodiversity.

There is a contradiction, therefore, in the global political-economic system needing biodiversity at the global level while it diminishes it at the local level. As Altieri and Merrick (1987, 361) write "Today, the foundation and health of agriculture in industrial countries largely depend on their access to the rich crop genetic diversity found in Third-World countries. Yet the very same germplasm resources most sought after for their potential application in biotechnology are constantly threatened by the spread of modern agriculture." This contradiction is, in turn, part of a more fundamental paradox. There is a tendency for political-economic centers to incorporate—to make the other the same as the self—even when such a transformation contravenes the center's own best interests. This tendency can be seen in the current interest in using the concept of intellectual property rights to "rationalize" the relationship between the more developed and the less developed countries with respect to indigenous knowledge, biodiversity, and community rights. This tendency can be attributed to the ideological inclination of political-economic centers to devalue their opposite, the periphery, and—especially—the centers' attendant disinclination to associate anything that they need at the periphery with the character per se of the periphery.

The position of communities at the margin of the global political-economic system is essential to their maintenance of biodiversity. To the extent that the global system (its biotechnology firms, health industry, human populations, and so on) needs this biodiversity, therefore, it also needs this marginality. At the same time that the global system needs marginality, however, it threatens it. This threat does not come solely from antagonism toward marginality: if the global system comes to value marginality, this valuation—and any efforts to manifest it—is likely to transform the marginality—to make the marginal less marginal. There is a paradox, therefore, in any effort by the global system to manage that which suffers from (or changes under) management. The paradox lies in the attempt to incorporate the role of the other while maintaining its otherness.

Incorporation

The paradoxical relationship between the periphery's value to and identity with the center can be further explored by examining the consequences of incorporation.

The Big Stone and the Small Man

There is a characteristic pattern of resource extraction by centers from peripheries in less developed countries with less equitable political-economic structures: whenever a resource at the periphery acquires value to the center, the center assumes control of it (e.g., by restricting local exploitation, granting exclusive licenses to corporate concessionaires, or establishing restrictive trade associations). The pattern is aptly expressed by a peasant homily from Kalimantan, which states that whenever a "little" man chances upon a "big" fortune he finds only trouble (Dove 1993b). He is in trouble because his political resources are not commensurate with his newfound economic resources. He does not have the power to protect and exploit great wealth, and so, inevitably, it is taken from him.

For the same reason, the proposed extension of intellectual property rights by the international community is not likely to reach such marginal people (or if it reaches them, it is not likely to stay with them) precisely because they are marginal. As Brush (1993, 665) writes "The history of prejudice and discrimination against indigenous peoples and ethnic minorities around the world suggests that the benefits from intellectual property protection for biological resources and indigenous knowledge will not reach indigenous people." Some proponents of the extension of intellectual property rights to peripheral communities worry that this may affect the lifestyles of these communities and thereby threaten their capacity to conserve biodiversity (e.g., Oldfield and Alcorn 1989, 206). I suggest that the real threat to their lifestyle and conservation capacity comes from the efficiency of resource flows not *into* but *out of* these communities. A defining characteristic of such communities is a differential in efficiency of resource flows, with flows out—as of biogenetic resources—being highly efficient and flows in—as of intellectual property rights compensation—being highly inefficient (Dove and Kammen 2001).

This same differential is reproduced at the international or global level (where it has drawn much more attention from scholars). Thus, Brush (1989b), in pointing out that the concept of "free" flows of resources tends to be applied only to flows from less to more developed countries, writes: "Advanced agricultural and industrial nations have stressed the need for free exchange of germplasm. Some less developed countries with large reservoirs of exotic germplasm have argued that the exchange of germplasm is "free" only because it flows from underdeveloped to developed regions" (21). Kloppenburg (1991, 16) suggests that this asymmetry is a product of a troubling differential in the way that the more developed countries value creation and production.

> Indigenous people have in effect been engaged in a massive program of foreign aid to the urban populations of the industrialized North. Genetic and cultural information has been produced and reproduced over the millennia by peasants and indigenous people. Yet ... the fruits of this work are given no value despite their recognized utility. On the other hand, when such information is processed and transformed in the developed nations, the realization of its value is enforced by legal and political mandates.
>
> (16)

An Example of Poverty Alleviation in Indonesia

I will draw on a review of poverty alleviation programs in Indonesia under former President Soeharto's "New Order" government to illustrate this argument about relations between the center and the periphery. Most scholars and planners working in the field of poverty alleviation in Indonesia insist that one of their main challenges is to "reach" the poor with resources and that the chief obstacle to this is the myriad (and purported) shortcomings of the poor themselves. However, the findings of a government review team (of which I was a part) suggested that the challenge was not getting resources *to* the poor but rather getting resources *past* the interested parties and structural obstacles that stand between the poor and the central government (Dewey et al. 1993). A related challenge was ensuring that this interaction with the government bureaucracy did not result in extracting more resources from the poor than it provides to them.

Most antipoverty programs in Indonesia required, for example, a *sumbangan* (contribution) from prospective participants (whether in cash, labor, or materials), ostensibly to assure the government of the participants' sincerity and enthusiasm for the program. This contribution often excluded the true poor from participation in antipoverty programs, however, simply because they could not afford to make it. Even if the poor could afford to make the contribution requested, it might exceed the benefits of participation, so that the net impact of the program on the poor was a cost as opposed to a benefit. In truth, many of the New Order government's antipoverty programs, and many of its development programs in general, did not improve but worsened the economic circumstances of the people they were intended to benefit —a finding that was supported by the crash of the Indonesian economy in the late 1990s. Thus, the transmigration program was designed to improve the lot of some of Java's poorest farmers, but its intensive agricultural model foundered on the poor soils of Indonesia's outer islands. The nuclear estate program was designed to develop commercial smallholdings, but its internal economics failed to guarantee either farmer subsistence or the repayment of government loans. And the forest concession program was intended to extract timber and promote regional development, but it instead resulted only in local environmental and social degradation.

The recurrent failure of government antipoverty efforts pointed to a structural contradiction: the resources devoted to poverty alleviation efforts were channeled through the same political-economic structure implicated in the creation of this poverty. In other words, the same system that impoverished people was trying to lift them out of poverty. If we think of the poor as the periphery of this system and the national government as the center, we can see that at the same time that the center was impoverishing the periphery it was proposing integration as the cure. Efforts to integrate the poor by providing them with resources tended to fail, because they reversed the actual causality involved: the poor must first be integrated before they can receive these resources. (For example, as is shown by the government insistence on *sumbangan*, the poor must first be capable of making a contribution to the program before they can share in its resources.)

Trying to reach the poor with government aid is like trying to reach peripheral communities with intellectual property rights. In either case, these efforts are rendered problematic by the structural position of the intended recipients. These efforts can succeed only if this structural position is first transformed. In the case of the poor, this transformation would do away with the basis of poverty (as opposed to just providing aid); and in the case of the peripheral communities it would do away with some of this marginality and thereby perhaps do away with some of the support for biodiversity conservation.

The Heisenberg Indeterminacy Principle

The experience of antipoverty programs in Indonesia and other countries suggests that there is a basic paradox in resource relations between centers and peripheries (recollect that most rural communities are located somewhere along a continuum that runs from center to periphery). Something analogous to the Heisenberg indeterminacy principle prevails here: resources from the center cannot reach the peripheral community except through changing it.[5] Indeed, one conclusion of our review of poverty in Indonesia was that to change the peripheral community *is* to reach it. This has direct implications for center-periphery relations involving resource conservation. It suggests that the center cannot intensify interaction with the periphery without changing it and if it does change it then associated desirable qualities such as biodiversity conservation may be imperiled. As Lohmann (1991, 100) writes of Thailand:

> Even if the more flagrant abuses of developers and bureaucrats could be checked, there is no reason to believe that industrial society could either assimilate or preserve within itself the conservation practices of an [indigenous] society. Those practices derive their significance and viability from being part of a web of other practices specific to that society. To transfer those practices or their accompanying theories into a society dominated by market economics would result in ... the practices and theories being transformed into practices indistinguishable from those well-established in industrial society.

The history of involvement of central authorities in resource exploitation at the periphery—turning again to Indonesia for examples—is instructive in this regard. This history reveals a recurring cycle of resource exploitation, which begins with a rise in prices of some resource at the periphery, which in turn attracts the attention of central government and business. This is followed by governmental efforts to "rationalize" resource exploitation, ostensibly out of concern for the sustainability of the resource and the welfare of the local communities. In most such cases, however, these efforts result in a shift in the control and benefit of the resource exploitation away from local communities to outside interests. The result is to degrade the resource and impoverish the local community. Rattan, gold, and edible birds' nests are just three recent examples from Indonesia, of many that could be cited, that fit this pattern (see Dove 1993b). Intensified involvement by the state center in resource exploitation inevitably seems to result in killing the "goose that laid the golden egg." This tendency is characteristic of a particular type of state (exemplified in particular by Indonesia's former New Order regime). As Bunker (1985, 240) writes:[6] "The discrepancy between the power to repress and the power to direct and coordinate is seen most clearly in the actions of the developmental state in a nation of unevenly developed regions with heterogenous social formations comprised of different modes of production and extraction."

This ill-fated pattern of resource regulation reflects, at least in part, a structural conflict between centers of political-economic power and centers of natural resource wealth and diversity. Where the former flourish—or have extended their reach—the latter suffers. Conversely, centers of resource conservation and diversity persevere in areas that are relatively remote (whether geographically or structurally or both) from centers of political economy. This inverse relationship means that the greatest centers of biodiversity will tend to be those least integrated into important political-economic centers. This, in turn, means that precisely where biodiversity is greatest central intervention will be most difficult and the likelihood of precipitating undesirable changes as a result of this intervention will be highest. It also means that where biodiversity has been degraded its restoration may entail not increasing but decreasing linkages with the global system. Lohmann (1991, 93) writes of Thailand: "It is significant that the restoration of biological diversity ... is typically not connected with attempts to find an industrial market for the products of biologically diverse areas. Rather, it is tied to conscious and creative attempts to *disengage* from the wider market economy, and to put it in a subordinate place in social and cultural life." This is not to suggest that diversity is always diminished by greater political and economic incorporation of local communities into the global system (see Brush 1989b, 1992a, 1992d; Brush, Corrales and Schmidt 1988). Nor is this to suggest that increasing the linkages of local communities to the global system is always done at the initiative of the latter. Communities at the periphery of the global system not infrequently intensify linkages on their own terms and for their own ends (Dove 1993a, 1994b).

This structural conflict pertains to center-periphery relations not just at the national but at the international level. Whereas the international academic and development community tends to think of itself as part of the solution, it is often in fact part of the problem. One example from an international development project will help to illustrate this point. The Farm Forestry Research and Development Project (known as F/FRED in forestry development circles) was an international effort funded by the U.S. Agency for International Development to support the productive integration of trees into farming systems in Asia. This integration of annual food and perennial tree crops superficially appeared to be supportive of biodiversity conservation, and indeed it was on this basis that the F/FRED project was cited as a model international effort to promote sustainable agriculture (Cohen, Alcorn, and Potter 1991). In fact, however, the initial activity under this project (in which I participated) consisted of a review of multipurpose tree species throughout Asia in order to select the three to five "most important" species on which the project would concentrate. It was argued that by having project participants throughout Asia concentrate on the same few species, projectwide research activities and administrative procedures would be facilitated. While the motivation for this concentration was perhaps understandable, its impact on biodiversity was clearly more negative than positive. Thus, even the most "enlightened" and applauded international projects may fall victim to the center-biased imperatives described here.

The concept of intervention in this context, as in all development contexts, needs to be problematized. Brush (1992c, 7; 1993, 666) notes that the global political-economic system that proposes to cure the problem of unrecognized indigenous rights and waning biodiversity is the same system that gave rise to the problem. Incorporation (in an asymmetrical manner) into national and international political-economic structures is responsible for the uncompensated transfer of genetic resources and the diminishment of these resources. The fact that these same structures propose to resolve these problems through further linkages is more than merely ironic. I suggest that such ironic proposals are themselves but a further, subtler step toward incorporation and that whatever empowerment they promise participants is likely to be offset by unintended but structurally inevitable and disempowering consequences.

The Problem

The first step toward constructively "problematizing" international intervention with respect to biodiversity and the rights of indigenous communities is to critique the way in which this intervention has been structured in the past and to offer recommendations for the way that it should be structured in the future.

Questions of Equity

The problems that the advocates of intellectual property rights propose to address are very real. A number of scholars have convincingly documented the inequity involved in the free transfer of genetic biological resources from the less developed countries to the more developed ones (Brush 1989b, 21; 1992b, 1618–20; Altieri 1989, 77; Fowler and Mooney 1990; Kloppenburg 1991; Kloppenburg and Kleinman 1987; Shiva 1993b). The existence of this inequity is not open to question; what is open to question is whether it represents the whole picture. The inequity and degradation that attend relations between the more and less developed countries are matched—and sometimes exceeded—by those that attend relations between center and periphery within many of the same countries. This also applies to relations among the less developed countries, as Colchester (1994, 52) demonstrated in his critique of timber extraction from Central America by Malaysian firms (see Sizer and Rice 1995). Colchester writes (1994, 52): "The growth of such nepotistic, transnational elites casts doubt on the continuing relevance of development reforms that focus exclusively on addressing North-South inequities.

As the Guyana experience is beginning to reveal, North-South equity may count for less than nothing, unless accompanied by social justice within the South, as well as the North."

Posey (1990, 15) calculated that "less than 0.001% [one-thousandth of 1 percent] of the profits from drugs that originated from traditional medicines have ever gone to the indigenous peoples who led researchers to them." What we also need to know, however, is what percentage of these profits have gone—or might go under proposed uses of intellectual property rights—not to these indigenous peoples but to the political-economic elites of their countries. We can get some idea of this percentage by looking at the case of timber: the percentage of profits from tropical forest logging that goes to the indigenous inhabitants of these forests is likely similar to Posey's figure in order of magnitude (and yet the extraction of timber entails far greater costs for these inhabitants, by degrading their ecosystems and thus the basis of their subsistence, than does the extraction of biogenetic resources).

The fact that Posey presents us with a figure for shares of profits from traditional medicines and not for resources such as timber (or minerals, oil, etc.) may be telling. It is politically possible to calculate the percentage of drug profits that corporations in the industrialized countries have returned to indigenous peoples; it is far more difficult to calculate the percentage of timber profits that corporations in less developed countries have returned to indigenous peoples. Research on the latter topic is simply not permitted by many governments in the developing world.

The Players

Because inequity and resource degradation are as ubiquitous on the global landscape as proposals to address them, it is instructive to ask who supports the proposals to use intellectual property rights, why, and why now? Much of the support seems to come from scholars and activists in the industrialized nations (with minor support coming from political-economic—including governmental—elites in less-developed countries).[7] Notably absent from this dialogue are the voices of indigenous peoples and activists in these less developed countries. Their voices are being heard in criticism of exploitation of local resources by outsiders, but the focus typically is on who is responsible and does not extend to the issue of compensation. The fact that local communities are not asking for compensation in the name of their countries, coupled with the fact that their national governments are not addressing the issue of compensating local communities as distinct from the state, provides the first of several cautions against uncritical adoption of the intellectual property rights approach.

As for political-economic elites in the less developed countries, they often support the use of intellectual property rights because it represents an opportunity for a North-South wealth transfer in their favor; it represents an opportunity not to conserve resources but to appropriate as much as possible of any value realized from these resources by foreigners. (Failure to distinguish between these two highly disparate goals is a key misunderstanding of the debate over intellectual property rights.) This alone does not explain the elite interest in intellectual property rights, however. Some interest stems from the role that intellectual property rights may play in broader North-South dialogues: they potentially provide elites in the less developed countries with a way to rebut criticism by the more developed countries of their countries' records on environmental degradation and human rights violations. By accusing the industrialized nations of extracting without compensation valuable resources from indigenous peoples in the less developed countries, and by blaming this injustice for all subsequent resource degradation, the critique is effectively turned back on the critics.[8]

Scholars and activists in the industrialized nations have often failed to appreciate the relevance of this dialogue to the proposed use of intellectual property rights. This typifies the lack of political insight that traditionally characterized the international conservationist and environmentalist movement. An example is seen in Blum's (1993, 42) comment that all Indonesia requires in order to follow Costa Rica's model (i.e., the Merck-INBio agreement to use international compensation for intellectual property rights to support national conservation efforts) is "an information management system and a group of parataxonomists." It would be more accurate to say that all Indonesia needed to follow this model is a different political-economic past, present, and future. Blum herself notes (1993:42) that the Merck-INBio model is tailored to Costa Rica's democratic political system and might be less appropriate for nations with different political traditions. The fact that the proposed use of intellectual property rights has been welcomed by some nations with nonrepresentative governments should itself raise warning flags.

Discussions of intellectual property rights among scholars and activists in the industrialized countries have focused on resource-related abuses by political-economic elites in the more-developed countries and the potential of intellectual property rights to correct these abuses and rectify inequities in North-South relations. Proponents of the use of intellectual property rights have also directed criticism at northern scholars accused of past, unrequited exploitation of indigenous intellectual property rights. There has been much less discussion of resource-related abuses by political-economic elites in the less developed countries and the possibility that intellectual property rights may just be used to worsen these abuses and further exacerbate socio-economic inequities in these countries. Postmodernist scholars have suggested that true critical scholarship should not jeopardize the "reality" of the other without also jeopardizing the reality of the critic.[9] Proponents of the use of intellectual property rights to compensate indigenous people who justify this use in terms of the purported sins of fellow elites are not jeopardizing their own professional reality; indeed, they may well be protecting it by drawing attention away from the sins of host country governments. This fact should be weighed in considering the direction given to any plan to use intellectual property rights.

Re-Framing the Problem

As suggested by this discussion of who is involved with intellectual property rights and why, I submit that one of the main problems with their proposed use is the assumption that the state and the people are coterminous and the corollary assumption that compensation made to the state will therefore benefit the people. Both assumptions are implicit in the writings of most supporters of the proposed use of intellectual property rights. In fact, as O'Riordan (1993) suggests, there is no necessary identity of purpose between national agencies and indigenous peoples in the developing world. What does this mean for proposals to use the concept of intellectual property rights to compensate indigenous communities in developing countries for their development and conservation of biological resources?

The discontinuity between national agencies and indigenous peoples means, first, that compensation for intellectual property rights is likely to go to the former as opposed to the latter. As Brush (1993, 664) writes (see also Brush 1989a, 80): "Few states accept the rights of indigenous groups *per se*, and even fewer can be expected to lend their political and legal apparatus to promote intellectual property rights of indigenous groups." Second, this discontinuity means that compensation for intellectual property rights is not likely to be used for the ends intended. Even the model Merck-INBio agreement specifies that only half of the compensation will be used for conservation efforts, and there is no guarantee that *any* of it will reach indigenous communities.[10] As O'Riordan (1993) writes: "Ethnoscience deserves its own power base, which is not necessarily supported by a project (viz., Merck/INBio) that combines communal profitability and capacity building by ecological research institutes, no matter how well-intentioned the deals are." Third, and finally, the discontinuity means that the proposed use of intellectual property rights may wind up contributing not to the support but to the degradation of biodiversity and the rights of local communities to recognition for developing and conserving it. Any compensation—including that proposed within the intellectual property rights approach—attracts the attention of the central state to the local community, and both resources and rights tend to suffer under state intervention and exploitation.

The discontinuity between national agencies and indigenous peoples is increasingly recognized in international development policy circles. Such discontinuity is seen as potentially inimical to both the conservation of local resources and the safeguarding of local human rights. There is, as a result, a growing overlap of interest between the international lobby for resource conservation and the international movement for local rights. Breckenridge (1992, 774) writes: "Local community empowerment has become a force for achieving the international goal of sustainability, and international oversight has become a means for ensuring the empowerment of local communities. A synergistic alliance of global and local interests that places sovereignty in a new context has formed."

An international view of sustainable resource use is emerging that links it to protection of indigenous rights. Breckenridge (759) also notes that "Increasingly, environmental organizations endorse local communities' 'rights' while indigenous peoples and other local communities adopt the language of international environmental protection, assuming the role of 'trustees' on behalf of a broader community." This view posits a division of interest between nation-states and international corporations, on the one hand, and local communities and international environmentalist and human rights activists on the other. According to Breckenridge (1992, 739): "The ideas of global environmental rights and local community rights are mutually reinforcing, and the convergence of the two perspectives consequently represents a powerful alliance for change that challenges current government decision-making and seeks solutions through international law." This view does *not* posit a division of interest between the local communities and international scholars and activists, as is suggested by some of the proponents of the use of intellectual property rights (e.g., Blum 1993, 18; Fowler 1993). For example, Posey, (1990, 15) writes: "Many anthropoligists, biologists, ecologists and other researchers will oppose IPR because they know that they too will have to drastically change their 'life-styles.' Income from published dissertations and other books, slides, magazine articles gramophone records, films and videos—all will have to include a percentage of the profits to the native subjects."

I suggest that any such opposition to local communities among anthropologists and other scholars is a red herring. It detracts attention from other and far more serious sources of opposition to local community interest. It also is anomalous: in the case of all other major resources (timber, minerals, and so on), indigenous communities compete with and typically lose out to regional, national, and international political-economic structures. It is far-fetched to suggest that threats to genetic biological resources come, instead, not from political-economic elites but from a minor branch of the international social sciences.

The Solution

We can now propose some answers to questions that have been raised regarding the use of intellectual property rights in particular and development in general.

Development Addition versus Subtraction

Development interventions are typically conceived as some type of "addition," based on the premise that underdevelopment is caused by some type of "absence." The proposal to provide compensation for intellectual property rights represents this sort of addition. A corollary premise is that the incorporation into wider political-economic structures that is necessary for the purpose of such an addition is beneficial, or at least not harmful.

There is a revealing dichotomy in the way allegiance to these premises varies around the world. They are fervently supported in most of the world's political-economic centers (the ones that do the incorporating) and are just as fervently rejected in many of the world's hinterlands and peripheries (the ones that become candidates for incorporation). There are large parts of the globe where any contact with central authority is thought to be, on balance, not in one's best interest, and so all such contact is feared and avoided—and in many cases with good cause. Appreciation of this stance requires something like a paradigm shift for those who have implicit faith in the virtues and benevolent intentions of centralized government.

I will illustrate the dimensions of this paradigm shift with reference to the approach that some members of the international development community have taken to the problem of deforestation in less developed countries. This approach is typified by the "rain forest crunch" strategy, which refers to the global marketing of nuts and other products originating in endangered forests through the support of international activists and "green" consumers. The premises of this strategy are that the forest is being cut by local forest dwellers, that they are doing this because they are poor, and that they will cease doing this if lucrative markets can be found for the products of standing forest. The problem with this approach (see Dove

1993b) is that it shifts attention away from the international community's own role as resource degrader and focuses only on its potential role as "helper." The international community needs to ask not just what it can do to help but what it must do to stop hurting. The cessation of most deforestation depends not simply on stimulating benevolent intervention by the international community but on halting existing predatory interventions and not initiating new ones.

The same critique of "helping" applies to the proposed use of intellectual property rights. A naive intervention by the international community using this mechanism, no matter how well intentioned, is likely to fall prey to the problems just discussed, in which case the best intervention may be to not intervene or to subtract rather than add. What form might such subtraction take? We can begin to answer this question by looking at the ways in which government policies in both the more developed and less developed nations are (however unintentionally) not supporting but injuring biodiversity conservation and community equity. Elimination of existing policies that are inimical to biodiversity and community rights would be a good first step (see Brush 1992b, 1627). An obvious target would be national programs that contain disincentives for the continued cultivation of multiple and traditional varieties of crop staples. A less obvious but equally important target would be those aspects of the international development culture (e.g., research policies at the International Rice Research Institute [IRRI] and the other Consultative Group on International Agricultural Research [CGIAR] centers) that indirectly have the same effect.

Intellectual Property Rights

In the end, the intention of this analysis is less to critique the proposed use of intellectual property rights than to place it in a broader political-economic, and developmental, context and to draw on this wider perspective to seek a better solution to the problems of biodiversity conservation and economic equity for indigenous communities. Although the proposed use of intellectual property rights may be problematic, their nonuse—the status quo—also is problematic. The intention here is not to say that any use of intellectual property rights will inevitably have deleterious consequences for people and resources but only to say that such uses *may* have deleterious consequences and indeed *will* have them unless a conscious effort is made to alter the outcome. The message, is short, is that there is an inherent bias in many existing structures of political power and resource use that is likely to cause most of the proposed uses of intellectual property rights to have either no impact or an impact that is the opposite of that intended. The message is that the development context in which these conservation efforts are set is not a "neutral background" (Lohmann 1991, 100–1).

It should not be impossible to counter this structural bias, however, so that the proposed use of intellectual property rights can fill its intended function. The best way to do this is to problematize this bias and make it a central part of any use of intellectual property rights. That is, the best solution may be to make explicit in any intervention using intellectual property rights, the fact that the intervention itself is problematic and to provide means for addressing this. This would entail, in part, making it explicit that the intention of such interventions is to tackle the equity issues that are raised in relations between the more and the less developed countries and *also* those that are raised in relations between the national and local levels within some of the less developed countries. This would entail, in short, explicitly addressing the question of how the national government and local community will divide compensation for intellectual property rights, and what measures will be taken to ensure that the portion assigned to the local community both reaches it and stays with it.

Afterword

The proposal to use the concept of intellectual property rights to compensate indigenous communities for their development and conservation of genetic biological resources is clearly well intentioned, but if we can conclude anything from the foregoing analysis it is that good intentions are not enough. Any intervention by a superordinate political-economic entity in the affairs of a local community is potentially problematic, and any introduction of new resources to a local community is potentially hazardous to the welfare of some or all of its residents, no matter how well intentioned. This is especially true when economic and political welfare are dichotomized, as is typically the case in such interventions. Provision of economic resources without the political resources to control them is inherently problematic. It is impossible to change a people's economic position—which the proposed use of intellectual property rights try to do—without also changing their political position—which the proposed use does not address.

The attempt to artificially separate economic from political goals is a perduring characteristic of development planning by the more developed countries and a common explanation for the failure of this planning in the less developed countries (Ferguson 1990). Proponents of the use of intellectual property rights have a particularly clear choice before them—whether to learn from this experience or repeat it.

Acknowledgments

My research on the Kantu' was carried out during 1974–76 with support from the National Science Foundation (grant no. GS–42605). Research on other groups in Kalimantan and elsewhere in Indonesia was carried out during six years of subsequent work in Java (1979–85), with periodic field trips to Borneo and other outer islands, with support from the Rockefeller and Ford Foundations and the East-West Center's Program on Environment. My research on poverty in Indonesia was carried out during 1993, with support from the United Nations Development Program, under the auspices of Indonesia's central planning ministry, Badan Perencanaan Pembangunan Nasional (BAPPENAS). More recent work in Indonesia and the Philippines has been supported by a grant from the John D. and Catherine T. MacArthur Foundation. For recent data on the Melaban Kantu', I also draw on a re-study of them carried out with my collaboration by Endah Sulistyawati on the faculty of the Agricultural Institute of Bogor, Indonesia. None of the aforementioned people or organizations necessarily agrees with the analysis presented here, for which the author alone is responsible.

Notes

1 Further data on the Kantu' swidden system may be found in Dove 1985b, on Kantu' ritual and ecology in Dove 1993c, and on Kantu' rubber exploitation in Dove 1993a.

2 For some key contributions to this literature, see Breckenridge 1992; Brush 1992c; Greaves 1994; King, Carlson, and Moran 1996; Kloppenburg 1991; Mays et al. 1996; and Posey 1990.

3 A corollary principle, critiqued by Parry (2000), is that this compensation can be based on a conception of biogenetic resources as stock/material resources versus informational resources.

4 A classic analysis of this relationship is Burling's (1965) work on hill swiddens versus lowland *padi* fields. Dove (1985a, 2003) has looked at the contrast between society in and out of the forest.

5 This principle, drawn from quantum mechanics, states that accurate observation of one variable necessarily has adverse effects on the observation of other variables. The principle states, that is, that observation—or interaction—entails costs.

6 This is, in turn, part of a more general pattern of resource degradation. Ludwig, Hilborn, and Walters (1993, 17) write, "Although there is considerable variation in detail, there is remarkable consistency in the history of resource exploitation: resources are inevitably exploited, often to the point of collapse or extinction."

7 For example, of the six contributors to the 1991 volume, *Biodiversity: Social and Ecological Perspectives* (Shiva et al. 1991), five are northern activists and one is the head of a southern research foundation; and of the twenty-five contributors to the 1994 volume *Intellectual Property Rights for Indigenous Peoples: A Source Book* (Greaves 1994), twenty-two are scholars, activists, and business people working in institutions in North America and the remaining three are Native Americans.

8 See Dove and Khan (1995) for an analysis of this critique and counter-critique in the wake of a devastating cyclone in Bangladesh.

9 For example, Rabinow (1986, 252) faults the focus of scholars on colonial regimes as opposed to contemporary academia. He writes:

> Contemporary academic proclamations of anti-colonialism, while admirable, are not the whole story. Neither Clifford nor any of the rest of us is writing in the late 1950s. His audiences are neither colonial officers nor those working under the aegis of colonial power. Our political field is more familiar: the academy in the 1980s. Hence, though not exactly false, situating the crisis of representation within the context of the rupture of decolonization is, given the way it is handled, basically beside the point.

10 As Blum (1993, 20) writes of the agreement:

> The royalties that INBio receives from Merck will be used to support the conservation of Costa Rica's biodiversity. Ten percent of the initial $1 million plus 50 percent of the royalties go directly to Costa Rica's Ministry of Natural Resources; the rest of the profits are used to preserve the environment at the discretion of INBio and its board of directors.

Bibliography

Abdoellah O.S. 1991. Homegardens in Java and their structure development. In *Tropical Homegardens*, edited by K. Landauer and M. Brazil, 69–79. Tokyo: United Nations University Press.

Abdoellah O.S., H. Karyono, H.Y. Isnawan, Hadikusumah, Hadyana, and Priyono. 1978. *The Structure of Homegardens in Desa Selajambe and Desa Pananjung*. Bandung: Institute of Ecology, Padjadjaran University. (In Indonesian).

Abdoellah, O.S., Parikesit, Budhi Gunawan, Nani Djuangsih, Tatang S. Erawan. 1997. Biodiversity condition and its maintenance in the upper Citarum river basin. In *The Conditions of Biodiversity Maintenance in Asia*, edited by M.R. Dove and P.E. Sajise, 1–16. Honolulu: East-West Center.

Adi Haji Taha. 1989. Archeological, prehistoric, protohistoric, and historic study of the Tembeling Valley, Pahang, West Malaysia. *Jurnal Arkeologi Malaysia* 2:47–69.

Agrawal, A. 1995. Dismantling the divide between indigenous and scientific knowledge. *Development and Change* 26:413–39.

Alcala, A.C. 1986. *Guide to Philippine Flora and Fauna*. Vol. 10: *Amphibians and Reptiles.* Manila: National Resources Management Center and the University of the Philippines.

Alcala, A.C., C.C. Custodio, A.C. Diesmos, and J.C.T. Gonzalez. 1995. List of amphibians of Mt. Makiling, Laguna Province, Philippines, with notes on their population status. Paper delivered at the Fourth Symposium on Philippine Vertebrates and the annual meeting of the Wildlife Conservation Society of the Philippines, University of the Philippines, Diliman, Quezon City, April.

Alcorn, J. 1984. Development policy, forests, and peasant farms: Reflections on Huastec-managed forests' contributions to commercial production and resource conservation. *Economic Botany* 38: 389–406.

Altieri, M.A. 1989. Rethinking crop genetic resource conservation: a view from the south. *Conservation Biology* 3(1):77–79.

Altieri, M.A., D.K. Letourneau, and J.R. Davis. 1983. Developing sustainable agroecosystems. *BioScience* 33(1):45–49.

Altieri, M.A., and L.C. Merrick. 1987. In situ conservation of crop genetic resources through maintenance of traditional farming systems. *Economic Botany* 41:86–96.

———. 1988. Agroecology and in situ conservation of native crop diversity in the Third World. In *Biodiversity*, edited by E.O. Wilson, 361–69. Washington, D.C.: National Academy Press.

Alvard, M. 1993. A test of the ecological noble savage hypothesis: Interspecific prey choice by neotropical hunters. *Human Ecology* 21:355–87.

Anderson, K. 1999. The fire, pruning, and coppice management of temperate ecosystems for basketry material by California Indian tribes. *Human Ecology* 27(1):79–113.

Anderson, R.S., E. Levy, and B.M. Morrison. 1991. *Rice Science and Development Politics: Research Strategies and IRRI Technologies Confront Asian Diversity (1950–1980)*. New York: Oxford University Press.

Anderson, W.P. 1976. *Weed Science Principles*. New York: West Publications.

Anon. 1971a. The need for the conservation of Taman Negara. *Malayan Nature Journal* 24:96–205.

———. 1971b. Taman Negara: Introduction. *Malayan Nature Journal* 24:113–14.

———. 1994. Village monograph of Sukapura village, Wangisagara village, and Ranca Kasumba village, 1991–1994. Manuscript in Wangisagara village office.

Ashton, P.S. 1981. The need for information regarding tree age and growth in tropical forests. In *Age and Growth Rate of Tropical Trees: New Directions for Research*, edited by F.H. Bormann and G. Berlyn. Yale University School of Forestry and Environmental Studies Bulletin No. 94: 3–6. New Haven: Yale University.

Ashton, P.S., T.J. Givnish, and S. Appanah. 1988. Staggered flowering in the Dipterocarpaceae: New insight into floral induction and the evolution of mast flowering in the aseasonal Tropics. *American Naturalist* 132:44.

Aumeeruddy, Y., and J. Bakels. 1994. Management of a sacred forest in Kerinci Valley, Central Sumatra: an example of conservation of biological diversity and its cultural basis. *Journal d'Agriculture Tropicale et de Botanique Appliquée* 36(2):39–65.

Bajracharya, E. 1993. Vegetable in the terraces: Influences of commercial agriculture on Ifugao culture, ecology, and landscape. Ph.D. diss., University of Hawaii, Honolulu.

Balasegaram, M. 1996. Living in awe of the dragon. *The Star*, October 20.

Balée, W. 1989. The culture of Amazonian forests. *Advances in Economic Botany* 7:129–58.

———. 1993. Indigenous transformations of Amazonian forests: An example from Maranao, Brazil. *L'Homme* 33(2–4):231–54.

———. 1994. *Footprints of the Forest: Ethnobotany—the Historical Ecology of Plant Utilization by an Amazonian People*. New York: Columbia University Press.

———. 1998. Historical ecology: Premises and postulates. In *Advances in Historical Ecology*, edited by W. Balée, 13–29. New York: Columbia University Press.

Balée, W., ed. 1997. *Advances in Historical Ecology*. New York: Columbia University Press.

Balen, B. van. 1995. The birds of the Kayan Mentarang proposed national park, Kalimantan, Indonesia: Distributional records and conservation. Report for Project Kayan Mentarang, WWF-Indonesia Programme, Jakarta.

Bandy, D.E., D.P. Garrity, and P.A. Sanchez. 1993. The worldwide problem of slash-and-burn agriculture. *Agroforestry Today* 5(3):1–6.

Barbault, R., and S. Sastrapradja. 1995. Generation, maintenance, and loss of biodiversity. In *Global Biodiversity Assessment*, edited by V.H. Heywood, 193–274. Cambridge: Cambridge University Press.

Barbehenn, K., J.P. Sumangil, and J.L. Libay. 1973. Rodents of the Philippine Croplands. *Philippine Agriculturist* 56: 217–42.

Barton, R.F. 1919. Ifugao law. *University of California Publications in American Archaeology and Ethnology* 15(1):1–186.

———. 1922. Ifugao economics. *University of California Publications in American Archaeology and Ethnology* 15(5):385–446.

———. 1938. *Philippine Pagans: The Autobiographies of Three Ifugaos*. London: George Routledge and Sons.

———. 1946. *The Religion of the Ifugaos*. American Anthropological Association Memoir no. 65. Menasha, WI: American Anthropological Association.

———. 1955. *The Mythology of the Ifugaos*. Memoirs of the American Folkore Society, no. 46. Philadelphia: American Folklore Society.

———. 1969. Ifugao economics. *American Archaeology and Ethnology* 15(5):385–446.

Bates, D., and T.K. Rudel. 2000. The political ecology of conserving tropical rain forest: A cross-national analysis. *Society and Natural Resources* 13(4):619–34.

Bellwood, Peter. 1985. *Prehistory of the Indo-Malaysian Archipelago*. London: Academic Press.

Benjamin, G. 1997. Issues in the ethnohistoy of Pahang. In *Pembangunan arkeologi pelancongan negeri Pahang,* edited by Nik Hassan Shumaimi bin Nik Abdul Rahman, Mohamed Moktar Abu Bakar, Ahmad Hakimi Khairuddin, and Jazamuddin Baharuddin, 82–212. Pekan, Malaysia: Muzium Pahang.

Berger, J. 1972. *Ways of Seeing*. London: Penguin.

Bisby, F.A. 1995. Characterization of biodiversity. In *Global Biodiversity Assessment*, edited by V.H. Heywood, 21–106. Cambridge: Cambridge University Press.

Blaikie, P. 1985. *The Political Economy of Soil Erosion in Developing Countries.* London: Longman.

Blaikie, P., and H. Brookfield. 1987. *Land Degradation and Society*. London: Methuen.

Blower, J.H., N. Wirawan, and R. Watling. 1981. Survey of flora and fauna in the Cagar Alam Kayan Mentarang. Report for the WWF-Indonesia Programme, Jakarta.

Blum, E. 1993. Making biodiversity conservation profitable: A case study of the Merck/INBio Agreement. *Environment* 35(4):17–20, 38–45.

Bodley, J.H. 1990. *Victims of Progress.* 3d ed. Mountain View, Calif.: Mayfield.

Bormann, F.H., and G. Berlyn, eds. 1981. *Age and Growth Rate of Tropical Trees: New Directions for Research.* School of Forestry and Environmental Studies Bulletin no. 94. New Haven: Yale University.

Boserup, E. 1965. *The Conditions of Agricultural Growth: The Economics of Agrarian Change under Population Pressure.* London: Allen and Unwin.

Botkin D.B., and M. Sobel. 1975. Stability in time-varying ecosystems. *American Naturalist* 109: 625–46.

Brechin, Steven R., Peter R. Wilshusen, Crystal L. Fortwangler, and Patrick C. West. 2002. Beyond the square wheel: Toward a more comprehensive understanding of biodiversity conservation as social and political process. *Society and Natural Resources* 15(1): 41–64.

Breckenridge, L.P. 1992. Protection of biological and cultural diversity: Emerging recognition of local community rights in ecosystems under international environmental law. *Tennessee Law Review* 59(4):735–85.

Brookfield, Harold. 1999. *Exploring Agrodiversity: Issues, Cases, and Methods in Biodiversity Conservation.* New York: Columbia University Press.

Brookfield, H., and Y. Bryon. 1993. *Southeast Asia's Environmental Future: The Search for Sustainability.* Tokyo: United Nations University Press.

Brookfield, H., L. Potter, and Y. Byron. 1996. *In Place of the Forest: Environmental And Socio-Economic Transformation in Borneo and the Eastern Malay Peninsula.* Tokyo: United Nations University Press.

Brookfield, Harold, and Michael Stocking. 1999. Agrodiversity: Definition, description and design. *Global Environmental Change* 9:77–80.

Brookfield, Harold, Christine Padoch, Helen Parsons, and Michael Stocking, eds. 2002. *Cultivating Biodiversity: Understanding, Analysing and Using Agricultural Diversity.* London: ITDG Publishing for United Nations University.

Brosius, J.P. 1986. River, forest and mountain: The Penan Gang landscape. *Sarawak Museum Journal* 36:173–84.

———. 1990. *After Duwagan: Deforestation, Succession, and Adaptation in Upland Luzon, Philippines.* Michigan Studies of South and Southeast Asia, no. 2. Ann Arbor: Center for South and Southeast Asian Studies, University of Michigan.

———. 1991. Foraging in tropical rain forests: The case of the Penan of Sarawak, East Malaysia (Borneo). *Human Ecology* 19(2):123–50.

———. 1997. Endangered forest, endangered people: Environmentalist representations of indigenous knowledge. *Human Ecology* 25: 47–69.

Brown, L. 1996. Guru Lecture. Delivered to the European Parliament, February 1.

Brush, S.B. 1989a. Rejoinder to Altieri. *Conservation Biology* 3(1):80–1.

———. 1989b. Rethinking crop genetic resource conservation. *Conservation Biology* 3(1):19–29.

———. 1991. A farmer-based approach to conserving crop germplasm. *Economic Botany* 45:153–66.

———. 1992a. Ethnoecology, biodiversity, and modernization in Andean potato agriculture. *Journal of Ethnobiology* 12(2):161–85.

———. 1992b. Farmers' rights and genetic conservation in traditional farming systems. *World Development* 20(11):1617–30.

———. 1992c. Intellectual property rights and the biological resources of traditional agriculture. Paper presented at the annual meeting of the American Anthropological Association, San Francisco.

———. 1992d. Reconsidering the green revolution: Diversity and stability in cradle areas of crop domestication. *Human Ecology* 20(2):145–67.

———. 1993. Indigenous knowledge of biological resources and intellectual property rights: The role of anthropology. *American Anthropologist* 95(3):653–71.

Brush, S.B., M.B. Corrales, and E. Schmidt. 1988. Agricultural development and maize diversity in Mexico. *Human Ecology* 16(3):307–28.

Bryant, R., and M. Parnwell, eds. 1996. *Environmental Politics in Southeast Asia.* London: Routledge.

Bunker, S.G. 1985. *Underdeveloping the Amazon: Extraction, Unequal Exchange, and the Failure of the Modern State.* Urbana: University of Illinois Press.

Burkhill, H.M. 1971. A plea for the inviolacy of Taman Negara. *Malayan Nature Journal* 24:206–9.

Burling, R. 1965. *Hill Farms and Padi Fields: Life in Mainland Southeast Asia.* Englewood Cliffs, N.J.: Prentice-Hall.

Caldecott, J.O. 1988. *Hunting and Wildlife Management in Sarawak.* Gland, Switzerland: IUCN Tropical Forest Programme.

———. 1992. Enhancing cooperation between local people and protected area projects: Principles and practices. Working Papers no. 33. Environment and Policy Institute, East–West Center, Honolulu.

Cameron, I., and D. Edge. 1979. *Scientific Images and Their Social Uses: An Introduction to the Concept of Scientism.* Boston: Butterworths.

Cant, R.G. 1973. *An Historical Geography of Pahang.* Monographs no. 4. Kuala Lumpur: Malaysian Branch of the Royal Asiatic Society.

Carandang, W.M., and N.R. Lawas. 1992. Assessment of farming systems in the Makiling Forest Reserve. Development Action Program–Environment and Resource Management Program, University of the Philippines, Los Baños, Laguna, Philippines.

Carey, I. 1970. The religion problems among the Orang Asli. *Journal of the Malayan Branch of the Royal Asiatic Society* 43:155–60.

Carpenter, R.A. 1990. Biophysical measurement of sustainable development. *Environmental Professional* 12:356–59.

Cassman, K.G., S.K. De Datta, D.C. Olk, J. Alcantara, M. Samson, J. Descalsota, and M. Dizon. 1995. Yield decline and the nitrogen economy of long-term experiments on continuous, irrigated rice systems in the tropics. In *Soil Management: Experimental Basis for Sustainability and Environmental Quality*, edited by R. Lal and B.A. Stewart. London: CRC Lewis.

Cassman, K.G., and P.L. Pingali. 1995a. Intensification of irrigated rice systems: Learning from the past to meet future challenges. *GeoJournal* 35(3):299–305.

———. 1995b. Extrapolating trends from long-term experiments to farmers fields: The case of irrigated rice systems in Asia. In *Agricultural Sustainability: Economic, Environmental and Statistical Considerations*, edited by V. Barnett, R. Payne, and R. Steinde, 63–83. London: John Wiley and Sons.

Consultative Group on International and Agricultural Research (CGIAR). 1993. *CGIAR Annual Report 1992*. Washington, D.C.: CGIAR.

———. 1996. What is the CGIAR? <http://www.cgiar.org:80/whatis.html>.

Christanty, L. 1990. Home gardens in tropical Asia, with special reference to Indonesia. In *Tropical Home Gardens*, edited by Dalam K. Landauer and M. Brazi, 9–20. Tokyo: United Nations University Press.

Clark, T.W. 2002. *The Policy Process: A Practical Guide for Natural Resource Professionals*. New Haven: Yale University Press.

Clay, J.W. 1988. *Indigenous Peoples and Tropical Forests.* Cambridge, Mass.: Cultural Survival.

Cleary, M., and P. Eaton, eds. 1992. *Borneo: Change and Development.* New York: Oxford University Press.

Cohen, J.I., J.B. Alcorn, and C.S. Potter. 1991. Utilization and conservation of genetic resources: International projects for sustainable agriculture. *Economic Botany* 45(2):190–99.

Colchester, M. 1994. The new sultans: Asian loggers move in on Guyana's forests. *The Ecologist* 24(2):45–52.

Cole, J. 1985. Gua Kepayang '84: Taman Negara cave expedition. *Malayan Naturalist* 32:2–4.

Collar, N.J., and P. Andrew. 1988. Birds to watch: The ICBP world checklist of threatened birds. ICBP Technical Publications no. 8. International Council for Bird Preservation, United Kingdom.

Collins, J.L. 1986. Smallholder settlement of tropical South America: The social causes of ecological destruction. *Human Organization* 45(1):1–10.

Collins, Wanda, and Calvin Qualset, eds. 1999. *Importance of Biodiversity in Agroecosystems.* Boca Raton: Lewis/CRC Press.

Conklin, H. 1957. *Hanunóo Agriculture: A Report on an Integral System of Shifting Cultivation in the Philippines.* Forestry Development Papers, no. 12. Rome: Food and Agriculture Organization.

———. 1959. Shifting cultivation and succession to grassland climax. *Proceedings of the Ninth Pacific Science Congress* 7:60–62.

———. 1961. The study of shifting cultivation. *Current Anthropology* 2(1):27–59.

———. 1967a. Ifugao ethnobotany 1905–1965: The Beyer-Merrill report in perspective. In *Studies in Philippine Anthropology in Honor of H. Otley Beyer,* edited by M.D. Zamora, 204–62. Quezon City: Alemar-Phoenix.

———. 1967b. Some aspects of ethnographic research in Ifugao. *Transactions of the New York Academy of Sciences,* 2d ser., 30:99–121.

———. 1968. *Ifugao Bibliography.* New Haven: Southeast Asian Studies, Yale University.

———. 1980. *Ethnographic Atlas of Ifugao: A Study of Environment, Culture, and Society in Northern Luzon.* New Haven: Yale University Press.

Corry, S. 1993a. *"Harvest Moonshine" Taking You for a Ride: A Critique of the "Rainforest Harvest"—Its Theory and Practice.* London: Survival International.

———. 1993b. The rainforest harvest: Who reaps the benefits? *The Ecologist* 23(4):148–53.

Coster, C. 1927. Zur anatomie und physiologie der zuwachszonen und jahresringbildung in den tropen. *Annales Jardin Botanique de Buitenzorg* 38:1–130.

Cronon, W. 1996. The trouble with wilderness: Or, getting back to the wrong nature. In *Uncommon Ground: Rethinking the Human Place in Nature,* edited by W. Cronon, 69–90. New York: Norton.

Cruz, R.V., H.A. Francisco, and C.S. Torres. 1991. Agroecosystem analysis of the Makiling Forest Reserve. Development Action Program–Environment and Resource Management Program, University of the Philippines, Los Baños, Laguna, Philippines.

Danielsen, F., D.S. Balete, T.D. Christensen, M. Heegard, O.F. Jacobsen, A. Jensen, T. Lund, and M.K. Poulsen. 1991. Conservation of biological diversity in the northern Sierra Madre mountains in Isabela and southern Cagayan Province, the Philippines. Department of Environment and Natural Resources, Manila; International Council for Bird Preservation, United Kingdom; and Danish Ornithological Society, Copenhagen.

Datta, M., and Harun. 1978. *Report on the Study of Livestock Fodder.* Bandung: Fakultas Peternakan Universitas Padjadjaran. In Indonesian.

De Boef, W. 1990. Changing Ifugao agriculture: Vegetables as alternative to rice. MS thesis. Wageningen University.

De Cola, E. 1989. Fractal analysis of a classified Landsat scene. *Photogrammetric Engineering and Remote Sensing* 5(5):601–10.

Denevan, W.M. 1992. The pristine myth: The landscape of the Americas in 1492. *Annals of the Association of American Geographers* 82(3): 369–85.

Dentan, R.K. 2002. Disreputable magicians. The Dark Destroyer and the Trickster Lord: Reflections on Semai religion and a possible common religious base in South and Southeast Asia. *Asian Anthropology* 1:153–94.

Dentan, R.K., K. Endicott, A. Gomes, and M.B. Hooker. 1997. *Malaysia and the Original People: A Case Study of the Impact of Development on Indigenous People.* Needham Heights, Mass.: Allyn and Bacon.

Dewy, Alice, Michael R. Dove, Dwi Retnandari, Loekman Soetrisno. 1993. Conceptual issues in framing and addressing the problem of poverty. In: *Review of Poverty-Alleviation Efforts in Indonesia, 1968–1993: Micro-Reality and Macro-Context*. Report of the United Nations Development Programme Mission to Bappenas (Central Planning Ministry, Indonesia).

Directorate General of Livestock Service. 1994. *Statistical Book on Livestock*. Jakarta: Biro Pusat Statistik.

Dixon, C. 1990. *Rural Development in the Third World*. London: Routledge.

Dizon, J.T. 1992. Socio-economic profile of farmers in the Makiling Forest Reserve. Development Action Program–Environment and Resource Management Program, University of the Philippines, Los Baños, Laguna, Philippines.

Donovan, D.G., and Puri, R.K. 1998. Are we losing sight of the tree for the forest? Matching institutional and indigenous technical knowledge—the case of gaharu (*Aquilaria* spp.). Typescript.

Douglas, M. 1970. *Natural Symbols*. New York: Vintage.

Dove, M.R. 1983. Theories of swidden agriculture, and the political economy of ignorance. *Agroforestry Systems* 1:85–99.

———. 1985a. The agroecological mythology of the Javanese and the political economy of Indonesia. *Indonesia* 36:1–36.

———. 1985b. *Swidden Agriculture in Indonesia: The Subsistence Strategies of the Kalimantan Kantu'*. Berlin: Mouton.

———. 1993a. Para rubber and swidden agriculture in Borneo: An exemplary adaptation to the ecology and economy of the tropical forest. *Economic Botany* 47(2):136–47.

———. 1993b. A revisionist view of tropical deforestation and development. *Environmental Conservation* 20(1):17–24, 56.

———. 1993c. Uncertainty, humility and adaptation in the tropical forest: The agricultural augury of the Kantu'. *Ethnology* 40(2): 145–67.

———. 1993d. The responses of Dayak and bearded pig to mast fruiting in Kalimantan: An analysis of nature-culture analogies. In *Tropical Forests, People, and Food: Biocultural Interactions and Applications to Development*, edited by C.M. Hladik, A. Hladik, O.F. Linares, H. Pagezy, A. Semple, and M. Hadley, 113–23. Man and the Biosphere Series, vol. 13. Paris/Carnforth: UNESCO/Parthenon Publishing.

———. 1994a. North-south relations, global warming, and the global system. Special issue: "Human impacts on the pre-industrial environment." *Chemosphere* 29(5):1063–77.

———. 1994b. The transition from native forest rubbers to *Hevea brasiliensis* (Euphorbiaceae) among tribal smallholders in Borneo. *Economic Botany* 48(4):382–96.

———. 1996. Center, periphery, and biodiversity: A paradox of governance and a developmental challenge. In *Intellectual Property Rights and Indigenous Knowledge*, edited by S.B. Brush and D. Stabinsky, 41–67. Washington, D.C.: Island Press.

———. 2003. Forest discourses in South and Southeast Asia: A comparison with global discourses. In: *Nature in the Global South: Environmental Projects in South and Southeast Asia*, edited by P. Greenough and A. Tsing, 103–23. Durham: Duke University Press.

Dove, M.R., and D.M. Kammen 2001. Vernacular models of development: An analysis of Indonesia under the "New Order." *World Development* 29(4):619–39.

Dove, M.R., and M.H. Khan. 1995. Disaster discourse: Competing constructions of calamity—The case of the May 1991 Bangladesh cyclone. *Population and Environment* 16(5):445–471.

Dove, M.R., and T. Nugroho. 1994. Review of culture and conservation 1991–1994: A sub-project. Typescript. Funded by the Ford Foundation and the World Wide Fund for Nature's Kayan Mentarang Nature Reserve Project in Kalimantan, Indonesia.

Dove, M.R., and P.E. Sajise, eds. 1997. *The Conditions of Biodiversity Maintenance in Asia: The Policy Linkages between Environmental Conservation and Sustainable Development.* Honolulu: East-West Center.

DWNP (Department of Wildlife and National Parks, Malaysia), MNS, MMA, and WWF. 1986. *Tourist development proposals for the Kuala Tahan area of Taman Negara. World Wildlife Fund Malaysia.* Report produced under Project MAL 97/86: Taman Negara Development Plan. Prepared for the Prime Minister of Malaysia.

DWNP Malaysia, MNS, MMA, and WWF. 1987. *Tourist Development Proposals for the Kuala Tahan Area of Taman Negara.* Kuala Lumpur: World Wildlife Fund Malaysia.

Dwyer, P.D. 1996. The invention of nature. In *Redefining Nature: Ecology, Culture, and Domestication*, edited by R.F. Ellen and K. Fukui, 157–86. Oxford: Berg.

Dwyer, P.D., and M. Minnegal. 1991. Hunting in a lowland, tropical rainforest: Towards a model of non-agricultural subsistence. *Human Ecology* 19(2):187–212.

Eastman, J.R. 1993. IDRISI update manual, version 4.1. Clark Labs for Cartographic Technology and Geographic Analysis, Clark University, Worcester, Mass.

———. 1996. IDRISI for Windows user's guide. Clark Labs for Cartographic Technology and Geographic Analysis, Clark University, Worcester, Mass.

Eghenter, C. 2000. What is *tana ulen* good for? Considerations on indigenous forest management, conservation and research in the interior of Indonesian Borneo. *Human Ecology* 28(3):331–57.

Ehrlich, P.R., and E.O. Wilson. 1993. Biodiversity studies: Science and public policy. In *Perspectives on Biodiversity: Case Studies of Genetic Resource Conservation and Development*, edited by C.S. Potter, J.I. Cohen, and D. Janczewski, 233–38. Washington, D.C.: American Association for the Advancement of Science Press.

Ellen, R.F. 1986. What Black Elk left unsaid: On the illusory images of green primativism. *Anthropology Today* 2(6):8–12.

———. 1999. Forest knowledge: Forest transformation. In *Transforming the Indonesian Uplands*, edited by Tania Murray Li, 131–57. London: Harwood Academic Publishers.

Endicott, K. 1979a. *Batek Negrito Religion: The Worldview and Rituals of a Hunting and Gathering People of Peninsular Malaysia*. Oxford: Clarendon.

———. 1979b. The impact of economic modernization on the Orang Asli (aborigines) of northern Peninsular Malaysia. In *Issues in Malaysian Development*, edited by J. Jackson and M. Rudner, 167–204. Singapore: Heinemann.

———. 1984. The economy of the Batek Negritos of Malaysia: A problem of alternatives. *Canberra Anthropology* 2:7–22.

———. 1988. Property, power, and conflict among the Batek of Malaysia: Annual and historical perspectives. *Research in Economic Anthropology* 6:29–52.

———. 2000. The Batek of Malaysia. In *Endangered Peoples of Southeast and East Asia: Struggles to Survive and Thrive*, edited by L. Sponsel, 101–22. Westport, Conn.: Greenwood.

Endicott, K., and K. Endicott. 1986. The question of hunter-gatherer territoriality: The case of the Batek of Malaysia. In *The Past and*

Future of !Kung Ethnography: Critical Reflections and Symbolic Perspectives, edited by M. Biesele, R. Gordon, and R. Lee, 137–62. Hamburg: Helmut Basuke Verlag.

Engstrom, M.D. 1993. The development of biodiversity survey methods in zoology. to the Department of Mammalogy, Royal Ontario Museum, and the Museum Zoologicum Bogoriense, Bogor, Indonesia.

Erawan, T.S., N. Djuangsih, M. Muchtar, H. Setiana, and L.S. Istanti. 1997. Community structure and diversity of fauna in the upper Citarum river basin, West Java, Indonesia. In *The Conditions of Biodiversity Maintenance in Asia,* edited by M.R. Dove and P.E. Sajise, 73–112. Honolulu: East-West Center.

Esteva, G. 1987. Regenerating people's space. *Alternatives* 12(1):125–52.

Eyzaguiree, P.B. 1986. The ecology of swidden agriculture and agrarian history in São Tomé. *Cahiers d'Études Africaines* 101–102: 113–29.

Fairhead, J., and M. Leach. 1996. *Misreading the African Landscape: Society and Ecology in a Forest-Savanna Mosaic*. Cambridge: Cambridge University Press.

———. 1995. False forest histories, complicit social analysis: Rethinking some West African environmental narratives. *World Development* 23(6):1023–36.

FAO. 1994. *International Code of Conduct for Plant Germplasm Collecting and Transfer*. Rome: Food and Agriculture Organization of the United Nations.

Feder, E. 1983. *Perverse Development.* Quezon City: Foundation for Nationalist Studies.

Ferguson, James. 1990. *The Anti-Politics Machine: "Development," Depoliticization and Bureaucratic Power in Lesotho.* Cambridge: Cambridge University Press.

Fittes, A.H., and R.K.M. Hay. 1981. *Environmental Physiology of Plants.* London: Academic Press.

Forbes, A.A. 1995. Heirs to the land: Mapping the future of the Malaku Barun. *Cultural Survival Quarterly* 14:6–8.

Forman, R.T., and M. Godron. 1986. *Landscape Ecology*. New York: John Wiley.

Fowler, C.S. 1993. Intellectual property rights: Some considerations for the AAA. *Anthropology Newsletter,* May, 32.

Fowler, C., and P. Mooney. 1990. *Shattering: Food, Politics, and the Loss of Genetic Diversity.* Tucson: University of Arizona Press.

Fox, J., D.M. Truong, A.T. Rambo, N.G. Tuyen, L.T. Cuc, S. Leisz. 2000. Shifting cultivation: A new old paradigm for managing tropical forests. *BioScience* 50(6):521–28.

Frake, C.O. 1996. Pleasant places, past times, and sheltered identity in rural East Anglia. In *Sense of Place*, edited by S. Feld and K. Basso, 229–57. Sante Fe: School of American Research.

Francisco, H.A. 1997. Biodiversity maintenance at the farm level: Socioeconomic determinants and community perception. In *The Conditions of Biodiversity Maintenance in Asia*, edited by M.R. Dove and P.E. Sajise, 267–80. Honolulu: East-West Center.

Francisco, H.A., F.K. Mallion, and Z.M. Sumalde. 1992. Economics of dominant forest-based cropping systems in Mount Makiling Forest Reserve: An integrative report. Development Action Program–Environment and Resource Management Program, University of the Philippines, Los Baños, Laguna, Philippines.

Frossard, D. 1994. Peasant science: Farmer research and Philippine rice development. Ph.D. diss., School of Social Sciences, Program in Social Relations, University of California, Irvine.

Fujisaka, S. 1986. Pioneer shifting cultivation, farmer knowledge, and an upland ecosystem: Co-evolution and system sustainability in Calminoe, Philippines. *Philippine Quarterly of Culture and Society* 14:137–64.

Ghimire, K., and M. Pimbert. 1997. Social change and conservation: An overview of issues and concepts. In *Social Change and Conservation: Environmental Politics and Impacts of National Parks and Protected Areas*, edited by K. Ghimire and M. Pimbert, 1–45. London: Earthscan.

Giddens, A. 1984. *The Constitution of Society: Outlines of the Theory of Structuralism.* Berkeley: University of California Press.

Gomez-Pompa, A., and A. Kaus. 1992. Taming the wilderness myth. *BioScience* 42(4):271–79.

Gonzalez, J.C.T. 1992. A comparative study on community structure: Birds of Puerto Galera, Mindoro Island, Philippines. Undergraduate thesis in zoology, University of the Philippines, Los Baños, Laguna.

———. 1993. An avifaunal survey of Puerto Galera, Oriental Mindoro Province, Philippines. *Asia Life Sciences* 2(2):163–76.

Gonzalez, J.C.T., and A.T. Dans. 1994. Microhabitat analysis of endemic frogs and skinks of Mt. Makiling, Laguna, Philippines. *Asia Life Sciences* 4(1).

———. 1997. Ecology and distribution of vertebrate fauna of Mount Makiling Forest Reserve. In *The Conditions of Biodiversity Maintenance in Asia,* edited by M.R. Dove and P.E. Sajise, 203–42. Honolulu: East-West Center.

Goodell, G. 1984. Untying the HYV Package: A Filipino farmer grapples with the new technology. *Journal of Peasant Studies* 11:238–66.

Gordon, R. 1990. The prospects for anthropological tourism in Bushmanland. *Cultural Survival Quarterly* 14:6–8.

Greaves, T., ed. 1994. *Intellectual Property Rights for Indigenous Peoples: A Source Book.* Oklahoma City: Society for Applied Anthropology.

Griffon, M. 1997. Towards a doubly green revolution. *Agricultural Rural Development* 4(2).

Grossman, L.S. 1984. *Peasants, Subsistence Ecology, and Development in the Highlands of Papua New Guinea.* Princeton: Princeton University Press.

Gruezo, W. 1990. Endangered plants of the Philippines. *Life Today* 46(8):16–18.

———. 1997. Floral diversity profile of Mount Makiling Forest Reserve, Luzon, Philippines. In *The Conditions of Biodiversity Maintenance in Asia,* edited M.R. Dove and P.E. Sajise, 153–202. Honolulu: East-West Center.

Gruezo, W., and J.C.T. Gonzalez. 1997. The functional relationships between plant and animal indicator species on Mount Makiling, Luzon, Philippines. In *The Conditions of Biodiversity Maintenance in Asia,* edited by M.R. Dove and P.E. Sajise, 243–66. Honolulu: East-West Center.

Guha, R. 1989. *The Unquiet Woods: Ecological Change and Peasant Resistance in Himalaya.* Berkeley: University of California Press.

Gunawan, B., O.S. Abdoellah, and E. Purnawan. 1997. The dilemma of dairy cattle project in upper Citarum river basin. In *The Conditions of Biodiversity Maintenance in Asia,* edited by M.R. Dove and P.E. Sajise, 113–24. Honolulu: East-West Center.

Guy, P., ed. 1995. *Agroecosystem Analysis and Rapid Rural Appraisal of Selected Sites in the Central Cordillera, Philippines.* ERMP Reports, no. 15. Environment and Resource Management Project, Philippines.

Hamilton, L S., and P N. King. 1983. *Tropical Forested Watersheds: Hydrologic and Soils Response to Major Uses or Conversions*. Boulder: Westview.

Hardjono, J. 1990. The dilemma of commercial vegetable production in West Java. Project Working Papers No. B-2, West Java Rural Nonfarm Sector Research Project, PSP-IPB, ISS, PPLH-ITB, Bandung, Indonesia.

Hargove, T. 1994. *A Dragon Lives Forever: War and Rice in Vietnam's Mekong Delta, 1969–1991, and Beyond*. New York: Ballantine.

Headland, T. 1997. Revisionism in ecological anthropology. *Current Anthropology* 38(4):605–30.

Heinen, J.T., and R.S. Low. 1992. Human behavioral ecology and environmental conservation. *Environmental Conservation* 19(2): 105–16.

Herdt, R.W., and C. Capule. 1983. *Adoption, Spread, and Production Impact of Modern Rice Varieties in Asia*. Los Baños: IRRI

Hirsch, P., and C. Warren, eds. 1998. *The Politics of Environment in Southeast Asia: Resources and Resistance*. New York: Routledge.

Hobart, M. 1993. Introduction: The growth of ignorance? In *An Anthropological Critique of Development: The Growth of Ignorance*, edited by M. Hobart, 1–30. London: Routledge.

Hughes, R., and F. Flintan. 2001. *Integrating Conservation and Development Experience: A Review and Bibliography of the ICDP Literature*. London: International Institute for Environment and Development.

Ingold, T. 1994. From trust to domination: An alternative view of human-animal relations. In *Animals and Humans Society: Changing Perspectives*, edited by A. Manning and J. Serpell, 1–22. London: Routledge.

IUCN (International Union for the Conservation of Nature and Natural Resources). 1994. *Red List of Threatened Animals*. Gland: Switzerland.

IRRI. 1991a. IRRI goal: A new rice plant type. *IRRI Reporter*, September.

———. 1991b. IRRI a major player in RF biotech network. *IRRI Reporter*, September.

———. 1992. *World Rice Statistics*. Los Baños, Philippines: IRRI.

———. 1993. *IRRI 1993–1994: Filling the World's Rice Bowl*. Los Baños, Philippines: IRRI.

Iskandar, J., and R. Kotanegara. 1993. *Methodology for Biodiversity Research.* Paper presented at the SUAN Biodiversity Methodology Workshop, Xishuangbanna, Yunnan, China, October.

Jabatan Hal-Ehwal Orang Asli Malaysia. 1961. *Statement of Policy Regarding the Administration of the Orang Asli of Peninsular Malaysia.* Kuala Lumpur: Department of Orang Asli Affairs.

Jackson, M.T. 1995. Protecting the heritage of rice biodiversity. *GeoJournal* 35(3):267–74.

Jong, W. de. 1995. Recreating the forest: Successful examples of ethnoconservation among Dayak groups in central West Kalimantan. In *Management of Tropical Forests: Towards an Integrated Perspective,* edited by O. Sandbukt, 295–304. Oslo: Center for Development and the Environment, University of Oslo.

Kalland, A. 2000. Indigenous environmental knowledge: Prospects and limitations. In *Indigenous Environmental Knowledge and its Transformations,* edited by R.F. Ellen, P. Parkes, and A. Bicker, 319–35. Amsterdam: Harwood.

Kantor Statistik Kabupaten Bandung. 1994. *Kabupaten Bandung Dalam Angka* [Statistics of the District of Bandung]. In Indonesian.

Kartawinata, K., S. Adisoemarto, S. Riswan, and A.P. Vayda. 1981. The impact of man on a tropical forest in Indonesia. *Ambio* 10:115–19.

Karyono. 1981. The structure of homegardens in rural areas of Citarum watershed, west Java. Ph.D. diss., Padjadjaran University, Bandung. In Indonesian.

Karyono, S.A., A. Ramlan, H. Ahmad, O.S. Abdoellah, Priyono, and I.S. Noer. 1978. *Ecological Study of Homegardens in Rural Areas of Citarum Watershed: The Structure of Homegardens.* Bandung: Institute of Ecology, Padjadjaran University. In Indonesian.

Kerkvliet, B.J.T. 1990. *Everyday Politics in the Philippines: Class and Status Relations in a Central Luzon Village.* Berkeley: University of California Press.

Khush, G.S. 1995. Modern varieties: Their real contribution to food supply and equity. *GeoJournal* 35(3):275–84.

King, S,R., T.J. Carlson, and K. Moran. 1996. Biological diversity, indigenous knowledge, drug discovery, and intellectual property rights. In *Intellectual Property Rights and Indigenous Knowledge,* edited by S.B. Brush and D. Stabinsky, 167–85. Washington D.C.: Island Press.

Kleinman, P.J.A., D. Pimentel, and R.B. Bryant. 1995. The ecological sustainability of slash-and-burn agriculture. *Agriculture, Ecosystems, and Environment* 52:235–49.

Klingman, G.C. 1982. *Weed Science: Principles and Practices*. New York: John Wiley.

Kloppenburg, J., Jr. 1991. No hunting! Biodiversity, indigenous rights, and scientific poaching. *Cultural Survival Quarterly* 15(3):14–18.

Kloppenburg, J. Jr., and D.L. Kleinman. 1987. The plant germplasm controversy: Analyzing empirically the distribution of the world's plant genetic resources. *Bioscience* 37(3):190–98.

Kruess, A., and T. Tscharntke. 1994. Habitat fragmentation, species loss, and biological control. *Science* 264(10 June):1581–85.

Lachowski, H. N.d. Guidelines for the use of digital imagery for vegetation mapping. Forest Service Remote Sensing Steering Committee, Integration of Remote Sensing in Nationwide Forestry Applications Program, Salt Lake City.

Lampe, K. 1995. Rice research: Food for 4 billion people. *GeoJournal* 35(3):235–59.

Landres, P.B., J. Vemer, and J.W. Thomas. 1988. Ecological uses of vertebrate indicator species: A critique. *Conservation Biology* 2(4):316–28.

Lawrence, D., D.R. Peart, and M. Leighton. 1998. The impact of shifting cultivation on a rainforest landscape in West Kalimantan: Spatial and temporal dynamics. *Landscape Ecology* 13:135–48.

Lawrence, D., and W.H. Schlesinger. 2001. Changes in soil phosphorus during 200 years of shifting cultivation in Indonesia. *Ecology* 82(10):2769–80.

Leaman, D.J., R. Yusuf, and H. Sangat-Roemantyo. 1991. *Kenyah Dayak Forest Medicines*. Jakarta: World Wide Fund for Nature Indonesia Programme.

Leman, L. 1987. *Report on analysis of land base and non-land-base dairy cattle business*. Bogor: Faperta Institut Pertanian Bogor. In Indonesian.

Lévi-Strauss, C. 1966. *The Savage Mind*. Chicago: University of Chicago Press.

Lewontin, R, 2002. The politics of science. *New York Review of Books* 49(8):28–31.

Meilleur, B.A. 1994. In search of "keystone societies." In *Eating on the Wild Side,* edited by N.L. Etkin, 259–79. Tucson: University of Arizona Press.

Miranda, H. 1987. Bird species diversity as related to the vegetation structure of disturbed lowland forests in Mt. Makiling. M.A. thesis, University of the Philippines, Los Baños, Laguna, Philippines.

Misliah bt. Mohd. Nashir, and Sahir b. Othman. 1996. Wildlife protection and management in Peninsular Malaysia. In *State of Environment in Malaysia,* edited by Consumers' Association of Pahang, 154–66. Penang: The Consumers' Association of Pahang.

Momberg, F., R.K. Puri, and T. Jessup. 2000. Exploitation of gaharu and conservation efforts in Kayan Mentarang National Park, East Kalimantan, Indonesia. In *People, Plants and Justice,* edited by C. Zerner, 259–84. New York: Columbia University Press.

Monmonier, M.S. 1984. Measures of pattern complexity for chloropleth maps. *The American Cartographer* 1(2):159–69.

Moran, E.F. 1993. Managing Amazonian variability with indigenous knowledge. In *Tropical Forests, People and Food: Biocultural Interactions and Applications to Development,* edited by C.M. Hladik, A.Hladik, O.F. Linares, H. Pagezy, A. Semple, and M. Hadley, 753–66. Man and the Biosphere, no. 13. Paris: UNESCO/Parthenon Publishing.

Myers, N. 1986. Tropical deforestation and a mega-extinction spasm. In *Conservation Biology: The Science of Scarcity and Diversity,* edited by M. Soulé, 394–409. Sunderland, Mass.: Sinauer Associates.

Nations, J.D. 1988. Deep ecology meets the developing world. In *Biodiversity,* edited by E.O. Wilson, 79–82. Washington, D.C.: National Academy Press.

Navarro, A.G. 1992. Altitudinal distribution of birds in the Sierra Madre del Sur, Guerrero, Mexico. *Condor* 94:29–39.

Needham, R. 1972. Punan-Penan. In *Ethnic Groups of Insular Southeast Asia,* Vol. 1: *Indonesia, Andaman Islands, and Madagascar,* edited by F.M. LeBar. New Haven: Human Relations Area Files Press.

Neumann, R. 1992. The political ecology of wildlife conservation in the Mt. Meru area of Northeast Tanzania. *Land Degradation and Rehabilitation* 3:85–98.

———. 1996. Dukes, earls and ersatz edens: Aristocratic nature and preservationists in colonial Africa. *Environment and Planning* 14: 76–98.

Nicholas, C. 2002. *The Orang Asli and the Contests for Resources: Indigenous Politics, Development, and Identity in Peninsular Malaysia.* IWGIA Documents, no. 95. Copenhagen: International Working Group for Indigenous Affairs.

Norgaard, R. 1987. The economics of biological diversity: Apologetics or theory? In *Sustainable Resource Development in the Third World,* edited by D.D. Southgate and J.F. Disinger, 95–109. Boulder: Westview.

———. 1989. Risk and its management in traditional and modern agroeconomic systems. In *Food and Farm: Current Debates and Policies,* edited by C. Gladwin and K. Truman, 199–216. Lanham, Md.: University Press of America for the Society for Applied Anthropology.

Norman-Myers, M.S. 1988. Environmental degradation and some economic consequences in the Philippines. *Environmental Conservation* 15(3):205–13.

Nyhus, Philip J., Frances R. Westley, Robert C. Lacy, and Philip S. Miller. 2001. A Role for Natural Resource Social Science in Biodiversity Risk Assessment. *Society and Natural Resources* 15:923–32.

O'Brien, T.G. 1998. Bulungan biodiversity survey. Preliminary report prepared for the Center for International Forestry Research, Bogor, Indonesia.

Oldfield, M.L., and J.B. Alcorn. 1989. Conservation of traditional agroecosystems: Can age-old farming practices conserve crop genetic resources? *BioScience* 37(3):199–208.

Olofson, H., ed. 1981. *Adaptive Strategies and Changes in Philippine Swidden Based Societies.* Quezon City: PDM Press.

Olson, S.L. and H.F. James. 1984. The role of Polynesians in the extinction of avifauna of the Hawaiian Islands. In *Quaternary Extinctions: A Prehistoric Revolution,* edited by P.S. Martin and R.G. Klein, 768–80. Tucson: University of Arizona Press.

O'Riordan, T. 1993. Editorial: Dealing in biodiversity. *Environment* 35(4):1.

Orlove, Benjamin S., and Stephen R. Brush. 1996. Anthropology and the conservation of biodiversity. *Annual Review of Anthropology* 25:329–52.

Padoch, C., and C. Peters. 1993. Managed forest gardens in West Kalimantan, Indonesia. In *Perspectives on Biodiversity: Case Studies of Genetic Resource Conservation and Development,* edited by C. Potter,

J. Cohen, and D. Janczewski, 167–76. Washington D.C.: American Association for the Advancement of Science.

Padoch, C., E. Harwell, and A. Susanto. 1998. Swidden, sawah, and in-between: Agricultural transformation in Borneo. *Human Ecology* 26(1):3–20.

Panayotou, T. 1994. Empirical tests and policy analysis of environmental degradation at different stages of economic development. *Pacific and Asian Journal of Energy* 4(1):23–42.

Pandey, U., and J.S. Singh. 1984. Energy-flow relationship between agro- and forest ecosystems in central Himalaya. *Environmental Conservation* 11(1):45–53.

Parikesit, D., and H. Yadikusumah. 1997. Spatial structure and floristic diversity of man-made ecosystems in upper Citarum river basin. In *The Conditions of Biodiversity Maintenance in Asia*, edited by M.R. Dove and P.E. Sajise, 17–44. Honolulu: East-West Center.

Parry, B. 2000. The fate of the collections: Social justice and the annexation of plant genetic resources. In *People, Plants, and Justice: The Politics of Nature Conservation*, edited by C. Zerner, 374–402. New York: Columbia University Press.

Peet, R., and M. Watts. 1996. Liberation ecology: Development, sustainability, and market in an age of market triumphalism. In *Liberation Ecologies: Environment, Development, Social Movements*, edited by R. Peet and M. Watts, 1–45. London: Routledge.

Pei Sheng-ji. 1991. Conservation of biological diversity in temple-yards and holy hills by the Dai ethnic minorities of China. *Ethnobotany* 3:27–35.

Peluso, N.L. 1992. *Rich Forests, Poor People: Resource Control and Resistance in Java*. Berkeley: University of California Press.

———. 1993. Coercing conservation: The politics of state resource control. *Global Environmental Change* 4(2):199–218.

Peluso, N.L., and C. Padoch. 1996. Changing resource rights in managed forests of West Kalimantan. In *Borneo in Transition: People, Forests, Conservation and Development*, edited by C. Padoch and N.L. Peluso, 121–36. Kuala Lumpur: Oxford University Press.

Pelzer, K.J. 1978. Swidden cultivation in Southeast Asia: Historical, ecological, and economic perspectives. In *Farmers in the Forest Economic Development and Marginal Agriculture in Northern Thailand*, edited by P. Kunstadter, E.C. Chapman, and S. Sabhasri, 271–86. Honolulu: University of Hawaii Press.

Peters, C.M. 1996. Beyond nomenclature and use: A review of ecological methods for ethnobotany. In *Selected Guideline for Ethnobotanical Research: A Field Manual*. Advances in Economic Botany, no. 10, edited by M.N. Alexiades, 24–166. New York: New York Botanical Garden.

Pickett, S., V. Parker, and P. Fiedler. 1992. The new paradigm in ecology: Implications for conservation biology above the species level. In *Conservation Biology: The Theory and Practice of Nature Conservation, Preservation, and Management*, edited by P. Fiedler and K. Jain, 65–88. New York: Chapman and Hall.

Pingali, P.L., P.F. Moya and L.E. Velasco. 1990. *The Post-Green Revolution Blues in Asian Rice Production: The Diminished Gap between Experiment Station and Farmer Yields*. IRRI Social Science Division Papers, no. 90–01. Manila: International Rice Research Institute.

Plotkin, M.J. 1993. *Tales of a Shaman's Apprentice: An Ethnobotanist Searches for New Medicines in The Amazon Rain Forest*. New York: Penguin.

Posey, D. 1985. Indigenous management of tropical forest ecosystems: The case of the Kayapo Indians of the Brazilian Amazon. *Agroforestry Systems* 3:139–58.

———. 1990. Intellectual property rights and just compensation for indigenous knowledge. *Anthropology Today* 6(4):13–16.

Posey, D., and W. Balée, eds. 1989. *Natural Resource Management by Indigenous and Folk Societies of Amazonia*. Advances in Economic Botany, no. 7. New York: New York Botanical Garden.

Potter, L. 2003. Forest versus agriculture: Colonial forest services, environmental ideas and regulations of land-use change in Southeast Asia. In *The Political Ecology of Tropical Forests in Southeast Asia: Historical Perspectives*, edited by Lye Tuck-Po, W. de Jong, and Ken-ichi Abe, 29–71. Kyoto: Kyoto University Press.

Power, A.G. 1996. Arthropod diversity in forest patches and agroecosystems of tropical landscapes. In *Forest Patches in Tropical Landscapes*, edited by John Schelhas and Russel Greenberg, 91–110. Washington, D.C.: Island Press.

Puri, R.K. 1992. Mammals and hunting on the Lurah River: Recommendations for management of faunal resources in the Cagar Alam Kayan Mentarang. Preliminary report for Project Kayan Mentarang, World Wide Fund for Nature Indonesia Programme, Jakarta.

———. 1995. Review of intraspecific prey choice by Amazonian hunters. By M. Alvard, *Current Anthropology* 36(5):809–10.

———. 1996. Post-abandonment ecology of settlement and activity sites in the Lurah River Valley, East Kalimantan, Indonesia. Quarterly research reports nos. 1–3. Indonesian Institute of Sciences. Jakarta, Indonesia.

———. 1997a. Hunting knowledge of the Penan Benalui of East Kalimantan, Indonesia. Ph.D. diss., Department of Anthropology, University of Hawai'i, Honolulu.

———. 1997b. Penan Benalui knowledge and use of treepalms. In *People and Plants of Kayan Mentarang*, edited by K.W. Sorensen and B. Morris, 194–226. London: World Wildlife Fund Indonesia Program/United Nations Educational, Scientific and Cultural Organization.

Rabinow, P. 1986. Representations are social facts: Modernity and post-modernity in anthropology. In *Writing Culture: The Poetics and Politics of Ethnography*, edited by J. Clifford and G.E. Marcus, 234–61. Berkeley: University of California Press.

Raeburn, P. 1989–90. Seeds of despair. *Issues in Science and Technology* 6(2):71–76.

Raffles, H. 2002. *In Amazonia: A Natural History*. Princeton: Princeton University Press.

Rambo, A.T. 1984. No free lunch: A re-examination of the energetic efficiency of swidden agriculture. In *An Introduction to Human Ecology Research on Agricultural Systems in Southeast Asia*, edited by A.T. Rambo and P.E. Sajise, 154–63. University of the Philippines, Los Baños, Laguna, Philippines.

Redford, K.H. 1990. The ecologically noble savage. *Orion Nature Quarterly* 9(3):25–29.

———. 1992. The empty forest. *BioScience* 42: 412–22.

Reijntjes, C., B. Haverkort, and A. Water-Bayers. 1992. *Farming for the Future: An Introduction to Low Input and Sustainable Agriculture*. Leiden: Macmillan.

Rickart, E.A., L.R. Heaney, and R.C.B. Utzurrum. 1991. Distribution and ecology of small mammals along an elevational transect on southeastern Luzon, Philippines. *Journal of Mammalogy* 72(3): 458–69

Robin, L. 2000. Yellowstone, edens and local people. Paper presented at the workshop Land Place Culture Identity, Institute of Advanced Studies, University of Western Australia, Nedlands, July 2–7.

Robinson, J., and K. Redford, eds. 1991. *Neotropical Wildlife Use and Conservation.* Chicago: University of Chicago Press.

Roosevelt, A.C. 1989. Resource management in Amazonia before the conquest: Beyond ethnographic projection. In *Resource Management in Amazonia: Indigenous and Folk Strategies*, edited by D.A. Posey and W. Balée, 30–62. Advances in Economic Botany no. 7. New York: New York Botanical Garden.

Roosevelt, A.C., et al. 1996. Paleoindian cave dwellers in the Amazon: The peopling of the Americas. *Science* 272: 373–84.

Rubeli, K. 1986. *Tropical Rain Forest in South-East Asia: A Pictorial Journey.* Kuala Lumpur: Tropical Press.

Saberwal, V.K. 1996. Pastoral politics: Gaddi grazing, degradation and biodiversity conservation in Himachal Pradesh, India. *Conservation Biology* 10: 741–49.

———. 1997. Saving the tiger: More money or less power? *Conservation Biology* 11:815–17.

Sahabat Alam Malaysia. 2001. *Malaysian Environment: Alert 2001.* Pulau Pinag: Sahabat Alam Malaysia.

Sajise, P.E., and N. Tapay. 1990. *UP National Assessment Project Report on the State of the Environment: Saving the Present for the Future.* Manila: University of the Philippines Press.

Sajise, P.E., and M.V. Ticsay-Ruscoe, with contributions from J.C.T. Gonzalez and William Sm. Gruezo. 1997. The conditions of biodiversity maintenance in Asia: Philippine studies. In *The Conditions of Biodiversity Maintenance in Asia,* edited by M.R. Dove and P.E. Sajise, 139–52. Honolulu: East-West Center.

Salazar, R. 1992. Community plant genetic resources management: Experiences in Southeast Asia. In *Growing Diversity: Genetic Resources and Local Food Security,* edited by D. Cooper, R. Vellvé, and H. Hobbelink. London: Intermediate Technology Publications.

Sanford, S.G. 1989. Crop residue/livestock relationship. In *Soil, Crop, and Water Management in the Sudano-Sahelian Zone: Proceedings of an International Workshop.* Niamey, Niger: International Crops Research Institute for the Semi-Arid Tropics Sahelian Center.

Saw, L.G., J.V. LaFrankie, K.M. Kochummen, and S.K. Yap. 1991. Fruit trees in a Malaysian rain forest. *Economic Botany* 45(1):120–36.

Sayogyo. 1988. Poverty problem in Indonesia: From theoretical and practical perspectives. *Mimbar Sosek Jurnal Sosial-Ekonomi Pertanian* 2:1–14. In Indonesian.

Schroth, Göth, et al. 2004. *Agroforestry and Biodiversity Conservation in Tropical Landscapes*. Washington, D.C.: Island Press.

Scoones, I. 1999. New ecology and the social sciences: What prospects for a fruitful engagement? *Annual Review of Anthropology* 28: 479–507.

Scott, C. 1996. Science for the West, myth for the rest? The case of James Bay Cree knowledge and construction. In *Naked Science: Anthropological Inquiry into Boundaries, Power, and Knowledge,* edited by L. Nader, 69–86. New York: Routledge.

Scott, J.C. 1985. *Weapons of the Weak: Everyday Forms of Peasant Resistance.* New Haven: Yale University Press.

———. 1998. *Seeing Like a State: How Certain Schemes to Improve the Human Condition Have Failed.* New Haven: Yale University Press.

Sellato, B.J.L. 1994. *Nomads Of The Borneo Rainforest: The Economics, Politics and Ideology of Settling Down.* Honolulu: University of Hawaii Press.

———. 1995. The Ngorek: A survey of lithic and megalithic traditions in the Bahau Area, East Kalimantan, and an interdisciplinary sketch of regional history. Report for the Culture and Conservation Project, World Wide Fund for Nature Indonesia Programme, Jakarta.

———. 2001. *Forest, Resources and People in Bulungan: Elements for a History of Settlement, Trade, and Social Dynamics in Borneo, 1880–2000.* Bogor, Indonesia: Center for International Forest Research.

Sheil, D., K.L. Puri, et al. 2002. *Exploring Biological Diversity, Environment, and Local People's Perspectives in Forest Landscapes: Methods for a Multidisciplinary Landscape Assessment.* Bogor, Indonesia: Center for International Forestry Research.

Shiva, V. 1988. Reductionist science as epistemological violence. In *Science, Hegemony, and Violence: A Requiem for Modernity,* edited by A. Nandy. New York: Oxford University Press.

———. 1991. *The Violence of the Green Revolution: Third World Agriculture, Ecology and Politics.* London: Zed Books.

———. 1993a. *Monocultures of the Mind: Perspectives on Biodiversity and Biotechnology.* London: Zed Books.

———. 1993b. The seed and the earth: Biotechnology and the colonisation of regeneration. *Canadian Woman Studies* 13(3):151–68.

———. 1994. Biodiversity conservation, people's knowledge, and intellectual property rights—an overview. In *Biodiversity Conservation: Whose Resource? Whose Knowledge?*, edited by V. Shiva. New Delhi: Indian National Trust for Art and Cultural Heritage.

Shiva, V., P. Anderson, H. Schucking, A. Gray, L. Lohmann, and D. Cooper. 1991. *Biodiversity: Social and Ecological Perspectives*. London and Penang: Zed Books and the World Rainforest Movement.

Siagian, S.P. 1988. *The management process of national development*. Jakarta: Gunung Agung. In Indonesian.

Sirait, M.T. 1997. The diversity of rattan and its uses: An example of natural resource management in Long Uli. In *The People and Plants of Kayan Mentarang*, edited by K.W. Sorensen and B. Morris, 181–93. London: World Wildlife Fund Indonesia Program and United Nations Educational, Scientific and Cultural Organization.

Sizer, N., and R. Rice. 1995. *Backs to the Wall in Surinam: Forest Policy in a Country in Crisis*. Washington, D.C.: World Resources Institute.

Soemarwoto, O., and Id. Soemarwoto. 1985. The Javanese rural ecosystem. In *An Introduction to Human Ecology Research on Agricultural Systems in Southeast Asia*, edited by T.A. Rambo and P.E. Sajise, 254–86. Los Baños: University Publications Program, University of the Philippines, Los Baños.

Soetomo, D. 1992. Sowing community seed banks in Indonesia. In *Growing Diversity: Genetic Resources and Local Food Security*, edited by D. Cooper, R. Vellvé, and H. Hobbelink. London: Intermediate Technology Publications.

Soesilo, Indroyono, H. Sanjaya, Prihartanto, and N.S. Heru. 1998. Bandung Basin is vulnerable to flood. *Kompas* April 3, 1998. In Indonesian.

Soliman, D. 1989. The challenge of participatory development: The MASIPAG experience. In *Proceedings of the Regional Workshop on Sustainable Agriculture*. Silang, Cavite, Philippines: International Institute for Rural Reconstruction.

Solnit, R. 1994. *Savage Dreams: A Journey into the Landscape Wars of the American West*. Berkeley: University of California Press.

Sorensen, K.W., and B. Morris, eds. 1997. *People and Plants of Kayan Mentarang*. London: World Wildlife Fund Indonesia Program and United Nations Educational, Scientific and Cultural Organization.

Soulé, M.E. 1991. Conservation: Tactics for a constant crisis. *Science* 253: 744–50.

———. 1995. The social siege of nature. In *Reinventing Nature? Responses to Postmodern Deconstruction,* edited by M.E. Soulé and G. Lease. Washington, D.C.: Island Press.

Sponsel, L.E. 1997a. The human niche in Amazonia: Explorations in ethnoprimatology. In *New World Primates: Ecology, Evolution, and Behavior,* edited by W.G. Kinzey, 143–65. New York: de Gruyter.

———. 1997b. The historical ecology of Thailand: Some explorations of thresholds of human environmental impact from prehistory to the present. In *Advances in Historical Ecology,* edited by W.A. Balée, 376–404. New York: Columbia University Press.

———. 2001. Is indigenous spiritual ecology a new fad? Reflections from the historical and spiritual ecology of Hawai`i. In *Indigenous Traditions and Ecology: The Interbeing of Cosmology and Community,* edited by J.A. Grim, 159–74. Cambridge, Mass.: Harvard University Press.

Sumalde, Z.M., H.A Francisco, and G.S. Fermintoza. 1992. Profitability analysis of dominant forest-based cropping systems in Makiling Forest Reserve. Development Action Program and the Environment and Resource Management Program, University of the Philippines, Los Baños, Laguna, Philippines.

Sumantri, M.A. 1984. *Petani Peternak Sapi Perah dan Pengembangannya* [Dairy cattle farming and its development]. Malang, Indonesia: Universitas Brawidjaya. In Indonesian.

Syahirsyah. 1997. Local knowledge in relation to secondary forest land and the utilization of forest resources. In *People and Plants of Kayan Mentarang,* edited by K.W. Sorensen and B. Morris, 101–12. London: World Wildlife Fund-Indonesia Program and United Nations Educational, Scientific and Cultural Organization.

Tan, B.C., E.S. Fernando, and J.P. Rojo. 1986. An updated list of endangered Philippine plants. *Taiwan* 3(2):1–5.

Terborgh, John. 1999. *Requiem for Nature.* Washington, D.C.: Island Press.

Thrupp, L.A., S. Hecht, and J. Browder. 1997. *The Diversity and Dynamics of Shifting Cultivation: Myths, Realities, and Policy Implications.* Washington, D.C.: World Resources Institute.

Tjitrosoepomo, G. 1988. *Plant Morphology.* Yogyakarta: Gadjah Mada University Press. In Indonesian.

Torres, C.S. 1992. Socio-economic factors affecting farm occupancy patterns in the Makiling forest reserve. Development Action Program–Environment and Resource Management Program, University of the Philippines, Los Baños, Laguna, Philippines.

Tuboh, L., E. Sipail, and Z. Gosungkit. 1999. A case study of the kadazan-dusun communities in the Crocker Range National Park, Sabah, Malaysia. In *From Principles to Practice: Indigenous Peoples and Protected Areas in South and Southeast Asia*, edited by M. Colchester and C. Erni, 206–19. IWGIS Documents, no. 97. Copenhagen: International Work Group for Indigenous Affairs and the Forest Peoples' Programme.

Turner, B.L. 1982. Pre-Columbian agriculture: Review of Maya subsistence. *Science* 217: 345–46.

Turner, B.L., W.C. Clark, R.W. Kates, J.F. Richards, J.T. Matthews, and W. Meyer, eds. 1990. *The Earth as Transformed by Human Action*. Cambridge: Cambridge University Press.

Valkenburg, J.L.C.H van. 1997. Non-timber forest products of East Kalimantan: Potentials for sustainable forest use. Ph.D. diss., University of Leiden.

Vayda, A.P. 1993. Ecosystems and human actions. In *Humans as Components of Ecosystems*, edited by M.J. McDonnell and S.T.A. Pickett, 61–71. New York: Springer-Verlag.

Vayda, A.P., T.C. Jessup, and C. Mackie. 1985. Shifting cultivation and patch dynamics in an upland forest in East Kalimantan, Indonesia. Final report to the U.S. Forest Service for the U.S.-Indonesian MAB Project, Jakarta.

Vickers, W. 1991. Hunting yields and game composition over ten years in an Amazon Indian territory. In *Neotropical Wildlife Use and Conservation*, edited by J. Robinson and K. Redford, 53–81. Chicago: University of Chicago Press.

Vom R. 1993. *The Migration of the Kenyah Badeng: A Study Based on Oral History*. Institute of Advanced Studies Monographs no. 7. Kuala Lumpur: Institute for Advanced Studies, Universiti Malaya. Partial translation of Roy, V. *The Kenyah Badeng*. Sibu, Sarawak: Society of Christian Service, 1988.

Wardle, D., M. Huston, J. Grime, F. Berendse, E. Garnier, W. Lauenroth, H. Setala, and S. Wilson. 2000. Biodiversity and ecosystem function: An issue in ecology. *Bulletin of the Ecological Society of America* 81(3):235–39.

WBL (Wildlife Biology Laboratory). 1994. Educational tour of Mt. Makiling Forest Reserve. An information pamphlet for Viajes Filipinas educational tours. Mount Makiling Forest Reserve: Wildlife Biology Laboratory.

Wells, M. 1994. Biodiversity conservation and local development aspirations: New priorities for the 1990s. In *Biodiversity Conservation: Problems and Policies,* edited by C. Perrings, K.C. Maler, C. Holling, C. Folke, and B.O Jansson. Boston: Kluwer Academic Press.

Wells, M., and K. Brandon. 1992. *People and Parks: Linking Protected Areas with Local Communities.* Washington D.C.: World Bank.

West, P.C. and S.R. Brechin, eds. 1991. *Resident Peoples and National Parks.* Tuscon: University of Arizona Press.

Western, D. 1989. Conservation without parks: Wildlife in the rural landscape. In *Conservation for the Twenty-first Century,* edited by D. Western and M.C. Pearl, 158–65. New York: Oxford University Press.

Western, D., and R.M. Wright, eds. 1994. *Natural Connections: Perspectives in Community-Based Conservation.* Washington D.C.: Island Press.

Wharton, Charles H. 1968. Man, fire and wild cattle in Southeast Asia. In *Annual Proceedings of the Tall Timbers Fire Ecology Conference,* edited by E.V. Komarek, 107–67. Tallahassee: Tall Timbers Research Station.

Whitten, T., D. Holmes, and K. MacKinnon. 2001. Conservation biology: A displacement behavior for academia? *Conservation Biology* 15(1):1–3.

Widagda, L.C., O.S. Abdoellah, G. Marten, and J. Iskandar. 1984. *Traditional Agroforestry in West Java: The Pekarangan (Homegarden) and Kebun-talun (Perennial-annual Rotation) Cropping System.* Honolulu: East-West Center.

Wildlife Commission of Malaya. 1932. *Report of the Wildlife Commission,* 3 vols. Singapore: Government Printing Office.

Williams, N., and E.S. Hunn, eds. 1982. *Resource Managers: North American and Australian Hunter-Gatherers.* Boulder: Westview.

Winarto, Yunita Triwardani. 2004. *Seeds of Knowledge: The Beginning of Integrated Pest Management in Java.* Monograph Series, no. 53. New Haven: Yale University Southeast Asia Studies.

Wood, D., and J.M. Lenné. 1999. *Agrobiodiversity: Characterization, Utilization and Management.* Wallingford (U.K.):CAB International.

World Bank. 1989. Indonesia: Forests, land, and water. Report no. 7822–IND, World Bank, Washington, D.C.

———. 1992. *The Development Report.* Washington, D.C: World Bank.

———. 1994. Indonesia environment and development: Challenges for the future. Report No. 12083–IND, World Bank, Washington, D.C.

Wulfraat, S., and Samsu. 2000. *An Overview of the Biodiversity of Kayan-Mentarang National Park.* Samarinda, Indonesia: WWF-Kayan Menterarang Project.

World Wide Fund for Nature, Malaysia. 1986. *The Development of Taman Negara.* World Wide Fund for Nature, Malaysia.

Yapa, L. 1993. What are improved seeds? An epistemology of the green revolution. *Economic Geography* 69(3):254–73.

———. 1996. Improved seeds and constructed scarcity. In *Liberation Ecologies: Environment, Development, Social Movements,* edited by R. Peet and M. Watts, 69–85. London: Routledge.

Yong, F. 1992. Environmental impact of tourism on Taman National Park in Malaysia. In *In Harmony with Nature, Malaysian Nature Journal Golden Jubilee Issues: Proceedings of the International Conference on Tropical Biodiversity*, edited by Yap Son Kheong and Lee Su Win, 579–91. Kuala Lumpur: Malayan Nature Society.

Zahid Emby. 1990. The Orang Asli regrouping scheme: Converting swiddeners to commercial farmers. In *Margins and Minorities: The Peripheral Areas and Peoples of Malaysia,* edited by V. King and M. Parnwell, 94–109. Hull: Hull University Press.

Zimmerer, K. 2000. The reworking of conservation geographies: Nonequilibrium landscapes and nature-society hybrids. *Annals of the American Association of Geographers* 90(2):356–369.

Index

Design, typography,
illustrations, and production
by **H.G. Salome** of

Hamden, Connecticut USA